S0-AAZ-525

$\mathcal{T}$he Northern Forest is one of the most resilient and valuable forests in the world. This huge, still largely unspoiled ecosystem lies within only a day's drive of the cities where one third of the U.S. population now lives.

Intense pressures are disrupting centuries-old patterns of land ownership and use, causing economic, ecological, and social upheaval. At the heart of these changes is a shift away from a view of the land as home and source of livelihood toward one that sees land as merely a commodity to be exploited for short-term profit.

Here's what readers have said about *The Northern Forest:*

"Astonishing. Wonderfully written, à la McPhee."

— Noel Perrin,
author of *First Person Rural* and *Last Person Rural*

"The Northern Forest is a very beautiful, sometimes strange and exotic place, as marvelous in its own way as the more celebrated tropical forests. What's going to happen to this land? Dobbs and Ober provide the answers in a timely, informative, and highly entertaining book."

—Ernest Hebert,
author of *The Dogs of March* and *Live Free or Die*

"Where *The Northern Forest* breaks new ground is in its exploration of . . . from the 'other' side: a morally embattled forester who knows damn well what kind of logging a healthy forest needs and can't do a thing about it if it means undermining double-digit quarterly reports; a hard-hustling independent logger barely eking out minimum wage; a disgruntled duck hunter who of his own initiative plants wild rice to improve waterfowl habitat, and never gets a word of thanks from the guys at Fish and Game . . ."

— *Orion*

"Dobbs and Ober have tapped the authentic voices of our region. This is the book I would hand to any visitor curious about us and say, 'Here's the life we lead here; this is the way it really is.'"

—Brendan J. Whittaker,
forester, clergyman, and Brunswick, Vermont selectman

THE NORTHERN FOREST

DAVID DOBBS & RICHARD OBER

Chelsea Green Publishing Company
White River Junction, Vermont
Totnes, England

Designed by Merrick Hamilton
Maps by Jill Shaffer
Set in 11 Point Adobe Garamond by Merrick Hamilton Book Design

Printed in the United States of America
00 01 02 03   3   4   5   6   7

First paperback printing February, 1996.

LIBRARY OF CONGRESS CATALOGING-IN-PUBLICATION DATA

Dobbs, David, 1958–
    The northern forest / David Dobbs and Richard Ober.
        p.   cm.
    Includes index.
    ISBN 0-930031-72-5 (cl)   0-930031-81-4 (pbk)
    1. Forests and forestry—New England. 2. Forests and forestry—
New York (State). 3. Forest management—New England. 4. Forest
management—New York (State). 5. Forest ecology—New England.
6. Forest ecology—New York (State). I. Ober, Richard. II. Title.
SD144.A12D63 1995
333.75'0974—dc20   95–6583

Chelsea Green Publishing Company
P.O. Box 428
White River Junction, Vermont 05001

CONTENTS

Acknowledgments ix
Introduction xiii

Part One NEW HAMPSHIRE 1
  1. Lake Umbagog 3
  2. Working for the Company 15
  3. Fragile Alliances 26
  4. Birding 36
  5. Lakeside 48

Part Two MAINE 59
  6. When a Tree Falls 61
  7. Imprints of the Past 73
  8. Chasing Chances 93
  9. The Industrial Forest 117
 10. Sleeping with the Elephants 137
 11. Continuum 149

Part Three VERMONT 161
 12. Cutting Maple 163
 13. A Farm's History 175
 14. Leafy Hollows and Pig Wallows 189
 15. Playing Santa 212
 16. Off the Farm 235
 17. Passing the Saw 249
 18. The End of the Season 258

Part Four NEW YORK AND BEYOND 265
 19. The Adirondacks 267
 20. Slippery Ground 299
 21. From the Ground Up 319

Glossary, Select Bibliography, Index 343

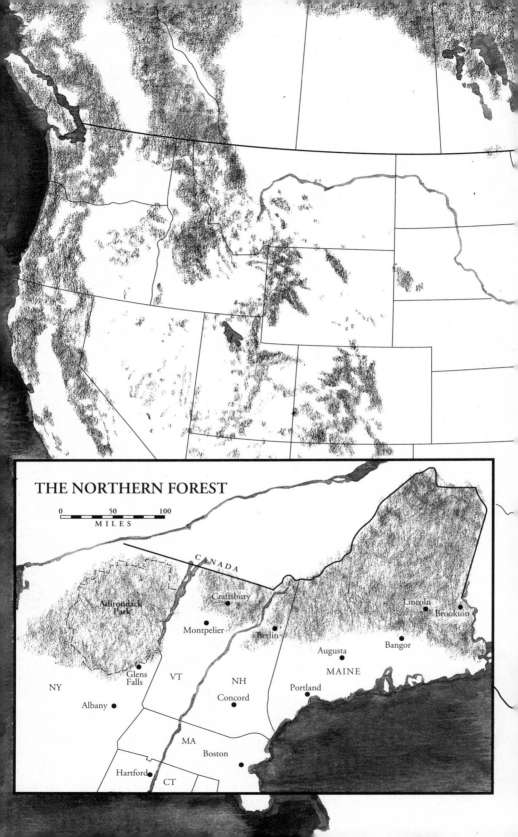

THE NORTHERN FOREST

0   50   100
MILES

CANADA

Adirondack Park

Craftsbury

Lincoln

Brookton

Montpelier

Berlin

Bangor

Augusta

MAINE

NY

Glens Falls

VT

NH

Portland

Albany

Concord

MA

Boston

Hartford

CT

FORESTS of the UNITED STATES

FOR CONNIE AND LIZ

ACKNOWLEDGMENTS

We owe unpayable debts to the many people whose help, cooperation, and support made this book possible, and whose interest and ideas sustained and cheered us through the last four years. The generosity and thoughtfulness we encountered is humbling to contemplate.

Our greatest debt is to the people who occupy most of the following pages: to Leo and Don Roberge of Berlin, New Hampshire; to Chuck Gadzik of Carroll, Maine; and to Jim, Joan, and Steve Moffatt of Craftsbury, Vermont. With nothing to gain but inconvenience, these individuals shared the details of their lives and their relationships with land, neighbors, and community, and did so with trust, honesty, and endless good humor. To encounter such openness is to find oneself faced with a great responsibility – that of answering to the trust and hopes of both subject and reader while remaining absolutely true to the material. We have striven with every sentence to do so, and can only hope that we have succeeded.

Several other people also shared with us their working and personal lives, taking the same risks as those who occupy greater parts of the book: In New Hampshire, biologists Jeff Fair and Carol Foss, and former Northern Forest Lands Council executive director Charles Levesque. In Maine, Pat Flood of International Paper; Dana Marble of Northeast Logging; Roger Milliken, Jr., of the Baskahegan Company; John Stewart of Champion International; Brian Souers, Dick Slike, and Andy Ward of Treeline, Inc.; and Dale and Jana Wheaton of Wheaton's Lodge. In Vermont, Michael and Penny Schmitt of the Inn on the Common in Craftsbury, and Larry Moffatt of Lyndonville. In New York, Richard Purdue, town supervisor of Indian Lake; Duane Ricketson of Olmsteadville; and Dale and Jeris French of Crown Point. Deep thanks to you all.

A different sort of debt, yet one equally vital, is owed to the scores of people who gave freely of their time, information, ideas, and opinions, either through conversations or through reading and commenting on drafts of part or all of the manuscript. Their contributions helped us refine our thinking and writing and greatly improved the book's accuracy and relevance; any errors remaining are strictly ours. In New Hampshire, we received such help from Eric Aldrich, Charles Baylies, Steve Blackmer, Paul Bofinger, Steve Breeser, Phil Bryce, Esther Cowles, Paul Doscher, Patrick Hackley, Chris Lincoln, Flip Nevers, Charles Niebling, Tudor Richards, Tammara Van Ryn, Jamie Sayen, Henry Swan, Sarah Thorne, and Brad Wyman; in Maine, Jerry Bley, Michael Cline, Lloyd Irland, Peter Ludwig, Sandy Neily, David C. Smith, and Bill Vale; in Vermont, Charles Cogbill, Joel Currier, Ken Davis, Bob De Geus, Jane Difley, Jim Esden, Charles Johnson, Mark McGrath, Bill Manning and everyone at the Citizens Forest Network, Peter Meyer, Stan Parsons, Jim Shallow, Steve Trombulak, and Brendan Whittaker; in New York, Susan Allen, Robert Bendick, Jim Frenette, Melinda Hadley, Jerry Pepper, Joe Rota, John Sheehan, and Neil Woodworth; and elsewhere, Adrienne Mayor, Josiah Ober, and Nina Ryan. We owe thanks to Tom Tuxill and Rudy Engholm of the Environmental Air Force for showing us the forest from the air, and to John Rosenberg and *Vermont* magazine for supporting this project in many ways.

We also each have several people to thank for more personal contributions.

We both wish to thank everyone at Chelsea Green Publishing Company for publishing authors and books the way they should be published, and for their unwavering support for this project. Ian Baldwin, our publisher, took an early and enthusiastic interest,

x

responded to all requests gracefully, and showed us every courtesy and consideration as we worked on the book. Jim Schley, our editor, who possesses not only a poet's ear and a sharp pencil but an all-too-rare combination of agile mind, alert eye, and flawless tact and wit, added immeasurably to the manuscript; in this day of phantom editors, we felt blessed to receive such intelligent and judicious attention. Peggy Robinson, marketing manager, and Suzanne Shepherd, production editor, both made contributions far beyond the call of their particular duties. Ben Watson also contributed invaluable criticism and ideas, firming up many soft spots in our thinking. Merrick Hamilton produced a wonderful design with speed, grace, and care. Ted Levin played an important early role by introducing us to Ian and encouraging us to pursue the project. And to N.P., who popped up unexpectedly like some wonderful surprise crop, we owe thanks for the finest sort of encouragement at the most critical of times.

From Richard Ober: I owe an immense debt of gratitude to my colleagues and friends at the Society for the Protection of New Hampshire Forests, not only for tolerating my sometimes erratic schedule over the past four years, but, more fundamentally, for teaching me about forests and people. Our time together, at work and in the woods and in innumerable conversations over the past decade, has done more than anything else to shape my conservation ethic. Special thanks in this regard go to Paul Bofinger, Rosemary Conroy, Geoffrey Jones, Ellen Pope, Sarah Thorne, and Tammara Van Ryn. As a professional, Charlie Niebling of the New Hampshire Timberland Owners Association answered my endless questions and challenged my biases; as a friend, he showed undying enthusiasm and always knew when to drag me into the woods for a ski. To my family – Nat and Marcia, Josh and Adrienne, Abby and Scott and Katy and Franklin, Susy and John, Ron and Jackie and Fred – your love and support is, quite simply, the foundation of my contribution to this book. It is hard to imagine a family more giving and kind. Finally, as with all things in my life, my

xi

work on this book was nurtured and strengthened by the love and creativity of my wife Liz. Of everything we share, I treasure most our commitment to work on that which matters.

From David Dobbs: Throughout the writing of this book I have been greatly helped by the support, encouragement, and love of my parents and siblings; their faith in me through the years has sustained me greatly, and their decency and strength keep me ever aware of the human capacity for good. I owe much also to Marylee MacDonald and John Wagner, both for their frank criticism and their endless, unconditional encouragement and support; every writer should have such friends. David Rawson and Karen Freedman have, through their love and immense goodness, given me strength, encouragement, and companionship vital to this undertaking. Likewise, I can only hope that David M. recognizes how much his support meant to me during the writing of this book. Finally, I have been sustained, through ways and means too various and voluminous to either enumerate or hope for, but for which I will always be grateful, by my wife Connie; I owe her my deepest thanks, and far, far more.

INTRODUCTION

This book is about one of America's last great forests and some of the people who live there. We could not write about one without the other, for the Northern Forest – 26 million acres stretching across upper New England and New York – is a landscape that has shaped its people as surely as they have shaped it.

We started working on this book in 1990, shortly after a sweeping effort to protect the Northern Forest had emerged as one of the Northeast's dominant environmental issues. When we first conceived the book, we planned to bring some local perspective to the debate by profiling a few residents. Before we got very far, however, we realized that the relationship between people and land was more than a good way to tell this complex story of the Northern Forest – it *was* the story.

From 1990 through 1994, we spent time with scores of people who live and work in northern Maine, New Hampshire, Vermont, and New York. We came to know several quite well. As these people showed us the woods and waters they call home, we were moved again and again by their care and respect for the land and for their communities, and by their understanding of the vital connections between the two. They sometimes expressed this understanding in words, but more often in simple actions: a forester kneeling in the soil to show us inch-tall spruce seedlings; a duck hunter searching the litter in a nesting box for signs of a successful hatch; a tree farmer carefully skidding maple logs between saplings that he will not live to harvest.

These relationships between people and land are among the rare qualities most imperiled by the forces that press upon the Northern Forest today. Excessive and short-sighted logging, development of remote and beautiful areas, and dwindling economic opportunities threaten to change the Northern Forest for the worse. Most of the people who live here recognize these pressures, yet they often feel powerless to influence the underlying causes of consumption and corporate priorities that drive them. At the same time, they feel isolated from the debate over how to solve such problems. In particular, they fear that public policies set in distant places, seats of government and politically powerful urban centers, will change their lives in no less dramatic fashion – that such policies, if not crafted with great care, will impose equally burdensome pressures on their lands and their lives.

The dilemma of the Northern Forest boils down to a single, vital question: Is it possible to protect a great forest without destroying the best parts of the resource-based economy and culture that both arise from and contribute to the land? This question resonates throughout the world; we've been given a unique opportunity to try to answer it here, in our country's first forest. Just as you can't understand the South American rain forest without considering the people who live there – the rubber tappers, the hunter-gatherers – you can't answer this question about the Northern Forest without knowing the woodcutters and mill workers, the foresters, the fishing guides, the small woodlot owners. They are what this book is about.

The region known as the Northern Forest spans 26 million acres, from the Atlantic coast of Maine to Lake Ontario and the St. Lawrence River in New York. It covers the upper two-thirds of Maine, the northern reaches of New Hampshire and Vermont, and the Adirondack and Tug Hill regions of New York. Encompassing one of the largest contiguously forested expanses in the nation, the Northern Forest of the United States is part of a larger bioregion that reaches north into the Canadian provinces of Quebec, Ontario, and New Brunswick, and west as far as Minnesota. The United Nations has declared approximately 11 million acres of this vast forest as an International Biosphere Reserve, the largest of its kind in the lower forty-eight states.

The distinctive characteristics of this region were shaped 12,000 years ago, when the mile-thick glaciers of the last ice age retreated, leaving scraped rock and a mix of glacial till – clay, sand, and gravel – that slowly broke down into soil. The forest that took root here occupies a broad transition zone where the northern hardwood communities of the eastern United States, dominated by maple, birch, and beech, mix with the boreal forests of spruce and balsam fir spreading down from Canada. This unique marriage of forest types, along with variations in terrain and geology ranging from sea-level peat bogs to alpine tundra above 5,000 feet, supports an unusually rich mix of life. The headwaters of the Northeast's primary river systems, including the Hudson, Connecticut, Merrimack, Penobscot, and St. John, all rise in this forest, which is drained by over 68,000 miles of streams and rivers altogether. The Northern Forest also contains more than seven thousand lakes covering nearly 1 million acres, and 2.5 million acres of wetlands. This variety of habitat is home to an estimated two hundred and fifty species of vertebrate wildlife, including twenty-five species unique to the region. Moose, bear, fox, and fisher make use of the uplands and wetlands, and dozens of species of fish and waterfowl thrive in the waters. The forest canopy and the diversity of wetlands, elevations, and terrain also provide nesting grounds for many of the hemisphere's migrant songbirds.

Approximately one million people live in the Northern Forest as well, mostly in small communities. Most live in clustered villages or in houses hidden away in the woods. Even the scattered commercial centers – the Plattsburghs, the St. Johnsburys, the Berlins and Bangors – are small, with only a few holding more than five thousand people. Overall, the population density averages about fifty people per square mile, or one-sixth what is typical of the twelve northeastern states; nearly half the Northern Forest has fewer than fifteen people per square mile. Vast reaches, especially in northern Maine, are virtually uninhabited.

Unlike the forests of the American West, which are largely publicly owned, most of the Northern Forest – 21.8 million acres, 84 percent of the total – is controlled by private landowners. Individuals own less than 40 percent of this private land, in tracts ranging from a few dozen to several thousand acres. The rest, more than 13 million acres, is owned in pieces of 5,000 acres or larger and managed for commercial forest products. Fewer than a dozen large paper corporations control 70 percent of this commercial forest land, with the rest held by roughly thirty-five sawmills, family-run firms, and other forest products investors. These private lands have supplied raw material for wood products manufacturing, which is the Northern Forest's dominant industry, since the late eighteenth century.

Ever since the Northern Forest was settled, some owners have treated the land well, while others have cut the forests hard. Cycles of exploitation have come and gone with waves of demand, innovations in harvesting and manufacturing, and forest regrowth. These cycles began soon after the first European colonists arrived. English settlers saw the New World, in the words of *Mayflower* passenger William Bradford, as a "hideous and desolate wilderness" that needed to be tamed. The Puritan colonists cut trees not only to make room for crops and to build shelter, but for spiritual salvation: Puritan culture held that forests were dark and gloomy and therefore bad, while fields were bright and airy and therefore good. Captain John Smith summed up this sentiment well in 1624. Exploring the wild Maine coast, he saw the untamed forest as "a Countrey rather to affright than delight one." He contrasted the frightful wilderness to "the Paradice" of the Massachusetts Bay Colony – paradise, because the trees were disappearing.

The settlers cut wood for another reason as well: profit. When the Pilgrims returned to England in 1621, they stacked their ship *Fortune* from hull to deck with pine clapboards. In wood-starved England, long deforested by the demands of agriculture, news of the colony's "infinite stores" of wood – especially its towering white pines – traveled fast. The Royal Navy paid special attention. For centuries, ship builders had made masts by splicing together sections of Scotch pine, a process that compromised both strength and resilience. With a single eastern white pine from the New World they could cut a perfect mast with wood to spare.

As the colonists expanded north from the Massachusetts coast, one of the first tasks in each settlement was to build a water-powered sawmill. The biggest and best trees were felled for buildings and furniture, and for making potash; the rest were cut for firewood and to clear fields. By 1635, when the first commercial sawmill opened in Maine, some New England families were measuring their wealth in clapboards and barrel staves. By 1800, wood dominated the export market from the region's seaports. Farmers and woodcutters had stripped entire watersheds in southern New Hampshire and southern Maine. Woodcutters had to search ever farther inland for the precious pine.

From the 1820s through the Civil War, the Northern Forest produced more timber than any place in the world. Wealthy families and timber companies from Boston, Portland, New York, and Philadelphia bought huge areas of remote forestland from the states and hired logging crews – some of them professional woodcutters (many French-speaking residents of Quebec), some of them farmers looking for off-season work – to spend the long winters cutting trees. Working out of primitive camps, wielding axes and crosscut saws, the loggers felled the best pine and spruce trees, cleaned off the branches, and dragged the logs with oxen and horses to the river banks. In the spring, the swollen rivers carried the logs downstream to water-powered sawmills.

By the 1850s, the landowners had cut virtually all of the tall softwood trees within hauling distance of the rivers. The "infinite stores" of wood were showing their limits. Then in the 1870s, logging railroads and portable steam-powered sawmills enabled the timber companies to reach deeper into previously inaccessible areas. These included the mountainous areas of all four Northern Forest states, as well as uncut forests that were far from the rivers. As the big trees in these last vestiges of the original forest were felled, the lumber boom collapsed. Logging companies moved south and west to find virgin woodlands. The farmers followed to find more fertile ground to till, or they moved to the cities to get manufacturing jobs.

With the big pine and spruce gone and the farm fields abandoned, thickets of hardwoods and smaller softwoods filled in the old cuts and clearings. Soon a second round of harvesting began as entrepreneurs developed new manufacturing processes to use this second-growth forest. For each successive species, new industries would swell up, often to collapse when the trees were gone; as new growth filled the void, so would another technology and another industry. Thus came and went the first long log mills, the tanneries, the clapboard mills, the boxboard factories.

Only the wood pulp and paper mills, which emerged in the 1880s, fully survived this period, for the paper industry needed only wood fiber to make its product, and could use trees of virtually any size. By 1910, paper companies had acquired huge areas of the Northern Forest from departing logging companies and farmers and had built mills throughout the region. These new companies included names familiar today, such as International Paper and Great Northern, as well as now defunct firms such as St. Regis and Brown. Papermaking rapidly became the region's dominant forest industry.

Meanwhile, before the turn of the century, heavy logging and resulting forest fires and erosion had given rise to the conservation movement. From 1880 through 1910, one wing of this movement, the preservationists, lobbied state and federal agencies to establish public forest reserves. The other wing, the utilitarians, advocated scientific forestry, tree planting, and fire control.

The conservationists won some important victories in the Northern Forest, preserving more than three million acres of land and inspiring some landowners to practice "sustained yield" forest management, which viewed forests as a renewable crop. Nevertheless, heavy cutting continued in many parts of the Northern Forest until the Great Depression, when lower demand for paper and increased production from commercial forests in the southeastern and western United States reduced harvest levels in the Northeast. From the 1930s to the 1960s, cutting trees to make paper and lumber remained the primary industry in the Northern Forest, but more careful forestry and lower investment in mills allowed much of the forest to grow back. Recreational use also increased during this period, as both visitors and residents used the backcountry lakes and forests for hunting, fishing, and camping. The large landowners leased out some land for exclusive use by individuals and clubs, but left most areas open to the public.

In the 1960s and 1970s, several multinational timber and paper companies, responding to growing demand for paper and armed with new tree-cutting and manufacturing technology, again stepped up cutting in the Northern Forest. Buying out some of the smaller firms, the new investors upgraded or replaced the region's aging paper and saw mills and built new plants to make "manufactured" building products such as waferboard. Driving this period of reinvestment were several factors: a flurry of road building that provided better access to remote areas; the development of new logging machines such as motorized skidders and whole-tree harvesters; and the rising value of the more mature forest. Logging increased through the late 1970s and early 1980s as the companies strained to feed the newer and faster mills.

To maximize return, many landowners of all sizes began to capitalize on the value of their land by expanding recreational leases and selling land to real estate speculators. Demand for vacation homes ran especially high during the economic growth period of the mid-1980s. It was then that a few prospectors, most notably Patten Corporation Northeast, began buying large chunks of the Northern Forest, subdividing them, and advertising "wilderness lots" in the real estate sections of the *New York Times* and the *Boston Globe.*

This trend toward wholesale subdivision alarmed a handful of veteran forestry and environmental activists, who in 1987 initiated a study by forest economist Perry Hagenstein. Hagenstein reported that the large commercial landowners that had come to control over half the Northern Forest were increasingly valuing some of their lands as real estate rather than as timber reserves, and he predicted that sales for development would continue. Though this concerned the conservationists who initiated the study, Hagenstein's paper attracted little attention otherwise.

Then in February 1988, a Boston-based timberland broker called LandVest confirmed Hagenstein's prediction with a dramatic announcement: the firm was offering for sale 186,000 acres of the Northern Forest. Scattered across northern New Hampshire, Vermont, and New York in parcels ranging from 40 to 40,000 acres, the land was owned by the European communications company Cie Générale Electricité (CGE). CGE had bought these properties, along with 790,000 acres in Maine, several months earlier from British financier and corporate raider Sir James Goldsmith. For Goldsmith, the land represented the bones of a fat bird called Diamond International Corporation, which he had caught in a hostile takeover in 1982 and systematically picked apart. Diamond had once used the land to feed its paper, lumber, and safety-match mills, but Goldsmith had carved off and sold these facilities to recover his initial investment. Selling Diamond's nearly one million acres to CGE was done purely for profit.

CGE clearly had no interest in managing timberland; the company was trying to cash in on the booming real estate market, and real estate speculators took notice. Within six months, two speculators bought all of the former Diamond lands in New York (96,000 acres), Vermont (23,000 acres), and New Hampshire (67,000 acres). They planned to subdivide much of the land into vacation lots, hotel or campground sites, private hunting leases, and ski developments. One of the buyers, Rancourt Associates, held a "Great North Country Land Auction" at the Mount Washington Hotel in New Hampshire to sell eighty lots – 30,000 acres in all – of the most developable Vermont and New Hampshire land. (About 12,000 acres were sold.)

CGE's 790,000 acres in Maine were controlled by a subsidiary called Diamond Occidental Forest, Inc. (DOFI). In 1988, DOFI sold 230,000 acres to Fraser Paper, and sold a 23% interest in the remaining land to James River Corporation, which owned mills in New Hampshire and Maine. Although James River and DOFI planned to keep much of the remaining 560,000 acres in timber management, they started analyzing the development potential of lakefront pieces. They also offered to sell land to conservation agencies, but public funds were scarce.

Few people at the time understood the details that had led up to the "Diamond" sale, as it became known, but many recognized that the transaction signaled a fundamental change in the Northern Forest: Land was becoming a commodity in and of itself, beyond its value for growing wood. Backcountry real estate was a corporate asset that could be sold as needed to improve cash flow, ward off a hostile takeover, or pay off debt.

Because half the region was owned by companies with profiles similar to Diamond's, environmentalists and others took this signal quite seriously. In the summer of 1988, conservation groups rushed to save some of the choice tracts that had been bought by developers. In New Hampshire, the Society for the Protection of New Hampshire Forests and The Nature Conservancy helped state and federal agencies secure 45,000 acres for public ownership. The State of New York protected more than half the Adirondack acreage through outright acquisition and conservation easements, which leave land in private hands but restrict development. Together, these deals cost the public over $15 million. Conservation groups, alarmed that more gigantic land sales would take place, told their members that a major conservation campaign was imminent.

Many leaders in the forest products industry were paying close attention as well. While not as anxious as the environmentalists, some industry leaders worried that large land sales could fragment the working forest, that new vacationers and second-home owners would object to logging, and that quality timberland would be removed from production. Others viewed the growing attention given to the Northern Forest as an opportunity to address tax and regulatory obstacles with which the industry had been grappling for years; the public might finally realize that growing wood for money is a tough business.

Among those alarmed by the Diamond sale were U.S. Senators Warren Rudman of New Hampshire and Patrick Leahy of Vermont. In 1988, the two senators shepherded a bill through Congress directing the U.S. Forest Service to conduct a comprehensive Northern Forest Lands Study. The governors of the four affected states appointed a special Governors Task Force on Northern Forest Lands to guide the work, which began in October 1988.

With the study underway, environmental groups turned up the heat. Many pointed out that the large landowners were not only selling land, but were also logging more heavily. The Wilderness Society called for creation of a 2.7-million-acre park in Maine. The National Audubon Society proclaimed that the debate over the Northern Forest would be even more difficult than the spotted owl conflict in the Pacific Northwest, which was just then reaching the crisis stage. Virtually every forestry, trade, and environmental group in the region turned its attention northward. When Maine's largest landowner, Great Northern-Nekoosa, which owned 2.1 million acres and one of the nation's biggest private hydroelectric systems, fell to a hostile takeover by Georgia-Pacific in February 1990, the transaction seemed to underscore the situation's urgency.

The Northern Forest Lands Study was completed in April 1990, incorporating research by two hundred specialists and the opinions of hundreds of residents from the four states. While the authors concluded that another attempt to liquidate an ownership as large as Diamond's was unlikely, they confirmed that smaller sales totaling tens of thousands of acres were fragmenting and degrading parts of the forest, especially along rivers and lakes. Further, the study found that large timberland owners, accustomed to earning annual returns of 6 percent from their best timberlands, were finding it hard to resist selling "nonstrategic" land (land that was too far from the company's mill, had been overcut, or contained less valuable tree species) for double-digit or even triple-digit returns. The authors predicted that continuing land sales and accompanying changes in use could kill jobs, degrade wildlife habitat, and block recreational access.

Congress had directed the Forest Service to avoid making specific recommendations, but in a "menu of alternatives" the study's authors suggested that better land-use planning, tax reform, public purchases of land and conservation easements, economic incentives for long-term forestry, and diversified economic development could help alleviate the pressures on the Northern Forest. The Governors Task Force endorsed these findings and concluded that the policies would require coordination across the region. The task force called for the formation of a second, high-level commission to further research these ideas, and to draft specific policy recommendations.

And so in 1990 the U.S. Congress and the four governors created the Northern Forest Lands Council (NFLC). The Council had seventeen members: four appointed by each of the state governors, representing industry, environmentalists, state governments and local interests, and one Forest Service official. By the time the Council began its work in 1991, the Northeast's real estate market had bottomed out, and the developers who had bought the Diamond lands were heading into bankruptcy. The underlying pressures that had driven the land sales still existed, however, and the public wanted to talk about these pressures. They wanted to talk about clearcutting and other logging practices, the effect of tax policies on private landowners and businesses, the loss of public access to private lands, declining logging and mill jobs, threats to biological diversity, poorly managed public lands, and most of all, the sense that the people of the region were losing control over their own destiny. For three years, the Council conducted a sort of regional town meeting involving thousands of people from across the Northeast. This was arguably the most exhaustive debate ever held over a regional land-use issue.

Meanwhile, the people of the Northern Forest continued to live their lives – some involved in the NFLC process, some ignorant of it; but many of them with a wary ear to the discussion, unsure how to participate.

The changes in land use and ownership pressing upon the Northern Forest are often described as "unprecedented." In one sense these changes are unprecedented, because demand for both wood and land has never been as high as it is now. But this characterization – particularly the implication that the forest was getting along fine in the decades before the 1980s – obscures the long history of cyclical exploitation and upheaval that began some 250 years ago.

Ironically, the cycles of heavy use that have periodically threatened the Northern Forest have been partly caused, or at least made possible, by the forest's remarkable resilience. Even in the face of rough treatment, these woods want to grow back, and they usually do. Some have cited this resilience to justify heavy cutting, and undoubtedly will continue to do so. But this rationale, which sees in the woods primarily the prospect of short-term gain, ignores the complexity and value of the forest's natural and human communities. This attitude worries many people for many reasons. It worries biologists who fear that heavy use will destroy the complexity that makes the forest so resilient in the first place. It worries wilderness advocates who recognize the biological and human benefits of untouched land. It worries forest economists who see greater long-term returns in growing bigger, older trees. And it worries loggers and mill workers who realize that if too many trees are cut down now, the trees may grow back, but not in their lifetimes.

These concerns sometimes inspire calls to "save" the Northern Forest by setting large portions of it aside. And there is a need for public preserves – to protect wildlife and genetic diversity, to maintain an ecological benchmark, and to provide recreation and solace for people. Yet there seems little question that most of the Northern Forest will remain a working forest. And it should. Until we solve the basic problems of population growth and American consumption, managing the Northern Forest for wood will remain an economic and moral imperative. As the most consumptive society in the history of the world, we cannot in good conscience set this forest aside and satisfy our appetite in other parts of the planet. We must learn instead to smooth out the waves of boom and bust, and to cultivate rather than exploit this forest's remarkable resilience.

Truly saving the Northern Forest will require setting aside certain areas and, more important, managing the rest with a much longer view. Consumers will have to pay more for some goods, so that prices reflect the true cost of good forestry. Industry and the environmental community will have to cooperate and make a commitment to the long term as never before, in politics as well as land use.

Most of all, truly saving the Northern Forest will mean recognizing and cultivating the region's other great resource – the knowledge and care residing in the people who live here, and who struggle each day to use these lands well. Most of these people care deeply about the land. They try to give back more than they take. Only if we understand, respect, and cultivate this relationship can we reach a lasting balance between ecology and economy – not just here, but in other places facing similar dilemmas around the world.

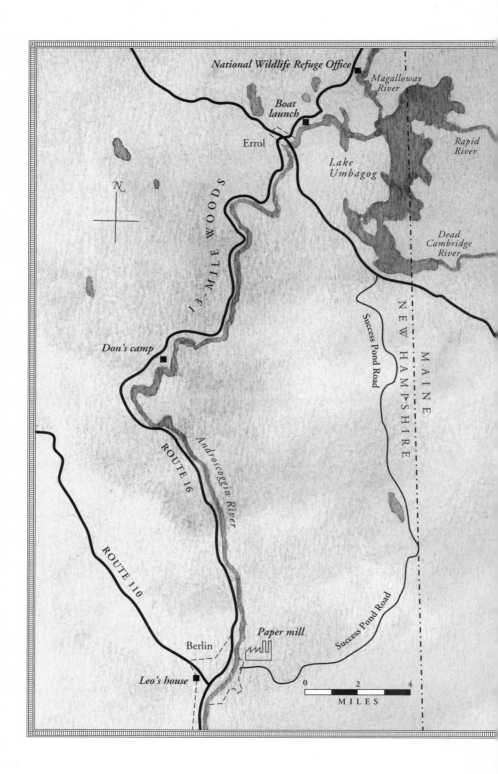

National Wildlife Refuge Office

Magalloway River

Boat launch

Errol

Rapid River

Lake Umbagog

13-MILE WOODS

Dead Cambridge River

N

Don's camp

Success Pond Road

NEW HAMPSHIRE

MAINE

ROUTE 16

Androscoggin River

ROUTE 110

Paper mill

Berlin

Success Pond Road

Leo's house

0   2   4
MILES

Part One

NEW HAMPSHIRE

—

We must be as courteous to a man as we are to a picture,
which we are willing to give the advantage of a good light.
RALPH WALDO EMERSON

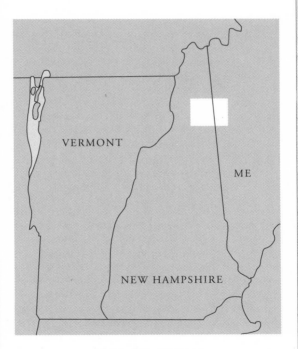

AREA OF DETAIL

I

Lake Umbagog

Lake Umbagog (Um-BAY-gog) is a great sprawling water that divides the spruce hills of northern New Hampshire from the low mountains of western Maine. The surrounding land is corrugated, rambling, thickly covered with the woods of the Northern Forest: the rounded crowns of maple, poplar, and birch trees provide a gauzy relief to the dark spires of spruce and balsam fir. The surrounding hills, rising in some places two thousand feet above the surface of the lake, may hold snow as early as September and as late as May.

Umbagog occupied only a thousand acres until 1851, when settlers in the town of Errol built a small dam at the lake's outlet to power a sawmill; later, timber companies enlarged and improved the dam – turned it into a spigot, essentially – so they could use the lake's water to float logs to mills downstream. The dam raised the lake's level only a few feet, but because Umbagog sat in a broad, flat basin, its waters eventually flooded an additional six thousand acres of low-lying forest and floodplain. Over the decades, these saturated lands developed into hundreds of acres of floating peat bogs, cedar and alder swamps, and riverine and lakeshore marshes.

Four rivers feed Umbagog. The Diamond empties a large watershed to the northwest; the Magalloway flows out of Parmachenee and Aziscoos Lakes to the northeast; the Rapid drains the huge Rangeley lakes area to the east; and the Cambridge delivers a more modest drainage from the southeast. A fifth river, the Androscoggin, emerges from the Umbagog dam to cut southward through New Hampshire's Coos (pronounced CO-oss) County. The waters of the

3

Androscoggin flow alternately through white-curled rapids and foam-flecked pools until the river reaches another dam at Pontook, 16 miles downstream, then runs free again for 14 miles to the city of Berlin and the neighboring town of Gorham, where it backs up behind several more dams.

The dams in Berlin and Gorham generate electricity to power the area's economic engine: the pulp and paper mill that sits on the Androscoggin's western bank in Berlin. To this mill flows water and wood from all over the Umbagog-Androscoggin watershed. The mile-long jumble of brick and metal buildings releases braids of steam and smoke from countless small vents and several huge smokestacks; it fills the air with sulfur's sharp tang. Conveyor belts, pipes, and tubes run from structure to structure. Logging trucks pull into the yard and stop before huge cranes with long dangling fingers; the cranes grab the logs and drop them onto piles two stories high. Great fans of water arc over the wood to keep it from drying out. At night the operation continues under bright lights, for the mill never rests.

As different as they are, the Androscoggin Valley's two defining features – clanging mill and placid lake – are connected by a long stream of entwined history. Both owe their existence to the Errol dam, which backed up the water that formed the lake and made possible the delivery of logs to the mill. This strange symbiosis has held since the late nineteenth century as the mill's owners, needing timber, kept the forest around the lake undeveloped. Thus the lake and its shores have fed the mill, and the mill, through its need for wood, both created the lake and helped protect it.

Roughly halfway between the lake at Umbagog and the mill at Berlin, squeezed into the strip of land between the Androscoggin River and New Hamphire Route 16, is a camp owned by Don Roberge and used almost every weekend by him and his brother Leo. Denise and Elaine, their wives, often come too, as do Don's kids, Melissa and Estacia, and Leo's, Keith and Mia. Don's is one of

4

several dozen "camps" built along this stretch of road; the structures run the spectrum from uninsulated, unplumbed shacks to fully winterized houses topped with satellite dishes. Don, when he bought the land in 1985, put a trailer on it – a little place to serve as a warm base for hunting and fishing expeditions. But Don and Leo have family and plenty of friends, and the trailer quickly proved too small. So one summer Don and Leo cut out one of the trailer's long walls and built to that side a large addition – wider than the trailer – into which they expanded the bath and kitchen and built a living room, a dining area, and a sleeping loft. To give the results of this work some architectural unity, the brothers covered both trailer and addition with vertical siding. But the place still didn't feel right. It seemed odd to have a trailer stuck into the side of what had become a house; the floors didn't quite mesh, and Leo and Don could feel the trailer there. So one weekend they chainsawed a large hole out of one end and towed the trailer away.

Now the floor is an unbroken plane and the place is quite comfortable. There's a couch and woodstove and TV in the living room and a stairway up to a sleeping loft; a hot dog cooker, a microwave, and a breakfast counter in the kitchen; and a new insulated porch with a bumper-pool table for the kids. The bathroom has a whirlpool tub. A garage in back is filled with gas tanks, life jackets, camouflage netting, cross-country skis, yard tools, snowshoes, ropes, a canoe.

Outside, a looped driveway provides parking for a boat and half a dozen vehicles. Even in winter, Don, Leo, and their friends spend a lot of time there, drinking coffee and beer, bouncing on their toes to keep warm, complaining about work, telling stories on each other, and reviewing and previewing woods outings. They wave to about half the vehicles that pass by. Sometimes a truck pulls in and the driver gets out to exchange reports on who-has-seen-what-where. These reports often lead to stories, the main difference between the two being that the stories, while generally true in their essentials, require the listener to suspend disbelief. You can tell that

5

reporting is shifting to storytelling when someone leans back against the grille of a truck, a foot on the bumper beneath, and says something such as, "Well, I remember when . . ." or "You know that time . . ." Because the men know the area around the lake well and have been on the lake, on the rivers, and in the surrounding forests with any number of friends and relatives uncountable times since they were boys, a story about a given person in a given place can easily lead to other stories involving that person or place, then branch off to stories about yet other persons and places, and so on, leading to the most surprising connections of people, places, and events. One afternoon during deer season, for instance, a story about flyfishing in the Androscoggin River evolved quite naturally from a story about a fistfight in a Berlin sewer.

On this particular November afternoon, it isn't long after the reporting has turned to storytelling that Don, acknowledging that realistically no one is going back out hunting again until they warm up and eat, nods toward the cabin. Inside, the woodstove is throwing heat, and the refrigerator holds the fruits of an earlier hunt. The men drape their hunter-orange vests and camo jackets over chairs near the stove to dry. With the temperature outside dropping through the teens, Don puts some moose stew on the electric range. Someone turns on the television, but the cartoons play to an empty couch. Everyone – Leo, Don, their cousin Rene, and their friends Marcel and Billy – is in the kitchen opening beer and preparing food.

Leo rips open a bag of chips and razzes Rene for blowing that morning's hunt. Leo, Rene, and Billy, communicating by walkie-talkie as they worked the woods on the other side of Route 16, had driven a deer into the sparse trees and regenerating remains of a large clearcut. "Right there," says Leo, "I had a shot. Deer comes across that knoll down in the cut and stops. And just as I'm sighting him in, Rene starts in on the radio, asking, 'Do you see him?' Deer didn't wait to hear the end of that! Soon as he heard those clicks, he went bing, bang, bong: gone. Had my rifle, I'd had him.

But I was hunting with the bow. The pin just went clink clink – through the trees." He laughs, thinking of the arrow caroming through the branches. "That sound we all know and love."

The men have been hunting with bow and arrow to increase the challenge, but with the season heading toward its end, they'll switch tomorrow to rifles. Laid out on a newspaper on the kitchen counter is a dismantled rifle and cleaning implements: ramrod, gun oil, felt swabs. Don moves it all aside to make room for a pot of moose stew and a six-pack of what he calls "nontoxic" – non-alcoholic – beer. "If anyone wants it," he says. Everyone else is already drinking Budweiser. Don opens a nontoxic; he has left his drinking days behind. The stories they tell, and the stories others tell about them, make it clear that these men have mellowed over the years, and that they had plenty of room in which to do so.

"Oh, Don, he's had some run-ins," says Leo, looking affectionately at his little brother. Three years separate them. "There used to be some rough times here. We had this motorcycle gang in town, the Road Kings. They couldn't fight their way out of a paper bag. I don't know why they thought they were so tough."

"They saw too many movies and started thinking they were the Angels," says Don, stirring the stew.

"They would pick up hitchhikers," Leo continues, "always kids you know, younger and smaller than themselves, and beat them up. These were just kids who hitchhiked because it was the neat way to get around then. They weren't hurting anybody. Finally the Road Kings beat up one of the kids we knew." Leo shook his head. "We don't like that too much. We got to set these guys straight. So one day Don's in town and he sees one of these guys climbing down the ladder into a manhole. He's going down to work on the sewer. Don goes over and yells down to the guy, 'Come up here.' The guys asks why. Don tells him, 'I want to beat the shit out of you for messing with those kids.' The guy says, 'Go away.' Don says, 'We can do it up here, or we can do it down there. It's no difference to me.'

"So the guy don't come up. Don goes down the ladder, beats him up right there in the sewer."

"Told him exactly why, too," says Don.

"Oh it was Don," says Leo, laughing. "It was always Don got in the shit. One time he's flyfishing in the Androscoggin. And if you're flyfishing over here, they can't go over there; they have to go where you are. The tourists, I mean. They gotta go right by you. Of course, you're flyfishing in the riffles, because that's where the fish are, but that's also where the canoeists want to go. But Don was there first. The first canoe goes right by him. Donnie gave 'em a few choice words. Next thing you know he got clapped behind the head by a paddle.

"He fell in the river. He was floating down and his friend pulled him out.

"Well, the shit hit the fan. After they got out of the water Don said, 'I know where they're going. I know those canoes.' And there they were at the canoe rental. He went in there, and he opened the door and said, 'One of those guys that rented your canoes just slapped me on the back of the head with a paddle.' They all laughed at him."

Leo, shaking his head, pauses to sip his beer.

"A bad thing to do," says Marcel.

"A bad thing to do," confirms Leo. "One guy got up, Don dropped him. The other three got up, he dropped them. Then he took a kayak and went through the place sticking the kayak through all the windows. Those guys were flying out of there then! Don cut his hand on one window, took fifty-five stitches to close his arm. He was bleeding like a stuck pig.

"Yeah, Don had to go to court for that one," says Leo. "Had to pay the fine. And he was just standing there flyfishing.

"But this kind of conflict happens more and more. You just do what you need to do to go about your business, and it creates ill taste with the out-of-towners. And most of the time they just don't understand. Like with the channels in the river. These people in

8

canoes, they stop right in the middle of the river to look at their birds or whatever. They'll be drifting along. Then you need to go near them in your motorboat because you have to follow this channel, and they give you this look. They don't even know where the channel is. They think you're just being rude. It's like they think we're aborigines or something.

"Me and Don were hunting a couple seasons ago next to the state park, and this lady from downstate's out there with her binoculars and her canoe, and she says, 'Don't you shoot my ducks.' We said 'Lady, these are not your ducks. They're just ducks that fly in. We can hunt them.' So she paddles off. She's not far when in comes a flight. Boom boom boom! We take a couple.

"She hit the top. Went to the warden and told him we were hunting in the state park. That was another one we had to go down for."

"Going down" means going to court. Don and Leo ended up paying $25 fines for that transgression, a fairly lenient penalty handed down by a judge who, after banging the gavel, winked and said, "There's ducks down in there, eh? Well, you boys stay out of my area."

It seems someone in the Roberge family is going down every few seasons. That same deer season, Don and Leo's father lost his deer hunting license when, catching up at the edge of town to a deer that he had wounded in the woods, and not wanting to fire a rifle close to town, he finished the deer off with his pistol – an act for which he didn't have a license.

The Roberges have had other run-ins with conservation law over the years. Asked about the brothers, wildlife officials – and some other people – tend to nod, subdue a smile, and say, "Ah, yes – the Roberge brothers." Leo and Don, says one official, "have been overzealous at times. They're avid hunters, and they've strayed over the line now and then. But we're talking about minor offenses." Most of the Roberges' transgressions were in their younger days – while the men were in their twenties and early thirties, before their

9

kids were grown, before Don stopped drinking – and many of the infractions seem quite literally to have been borderline transgressions: hunting too close to the state park, or on posted land.

It was Leo's incursion on to posted land, which occurred in 1989, that defined for him the changes creeping across the region. He and Marcel, out hunting geese, received a police escort off a piece of waterfront property they had hunted all their lives. When they got back to the road the cop showed them a NO TRESPASSING sign and asked if they could read. Leo said, Sure he could read, pointed out that the sign was a new one, and told the cop the place had never been posted in the thirty years he had hunted it. "Well," the cop said, "it's posted now. New owners."

"People from Massachusetts," says Leo. "They sell their homes for two hundred and fifty thousand bucks and come up here, they can buy a bunch of acres and a house for a hundred thousand and it seems like a great deal. Which to them I guess it is. But then the first thing they do after they buy, is they post it. That's not right. We've been going on this land all our lives, and we can't go there anymore."

What Leo was running into, of course, was the growth in second-home development that New England's real estate boom had brought to much of the Northern Forest in the 1980s. He had been seeing evidence of it for a couple of years by then, as FOR SALE signs went up on back roads and new owners built homes and posted their land against trespassing. Others in town had seen the same thing; the local pilot, who for years had earned his fees flying foresters, wildlife biologists, and the occasional tourist over the area, was suddenly flying a lot of land speculators around.

This increased development interest was not lost on the state's conservation groups, which for years had eyed Umbagog as a possible wildlife refuge. The lake's rich mosaic of wetland, riverine, and upland habitat had long made it a significant wildlife resource, and as most of New England's other large lakes had been developed over the decades, Umbagog's importance to wildlife had only increased. As early as the 1960s, Umbagog was one of the most valuable wildlife areas in all of northern New England. The lake and the land and rivers around it provided vital habitat for white-tailed deer, moose, black bear, coyote, fox, snowshoe hare, fisher, river otter, mink, muskrat, beaver, and countless smaller mammals; pickerel, perch, trout, and landlocked salmon; and songbirds that bred in both the upland woods and the shoreline tangles of spruce and alder. The lake was particularly important to waterfowl – both cavity nesters such as wood ducks, goldeneyes, and mergansers, which nested in the hollows of the many dead trees that poked up from the flooded wetlands and saturated shores; and ground nesters, including loons, mallards, ringed-neck ducks, and black ducks, that needed the protection provided by the brushy, secluded shoreline. Raptors too found the lake hospitable, and the area played a small but important role in New England's resurgence of osprey in the 1980s.

Then, in 1988, a pair of bald eagles began nesting on an island near the lake's western edge, in a small inlet called Leonard Pond. The eagles were one of the primary reasons cited by the U.S. Fish and Wildlife Service in 1990 when the agency proposed establishing the Lake Umbagog National Wildlife Refuge. The refuge idea grew out of concern among the region's environmentalists that the shores and hills around Umbagog, like so much other prime recreational lands in the Northern Forest, would be sold, subdivided, and developed. And indeed the three timber companies that owned most of the land around the lake were getting offers from private land speculators. After much persuasion from conservationists, the major landowner, James River Corporation, which owned the paper mill in Berlin, agreed to sell much of Umbagog's surrounding land to the government instead. The two other large owners later followed suit.

Like many area residents, Leo reacted to the refuge proposal with mixed emotions. On one hand, he recognized that the timber companies were under increasing pressure to sell; on the other, he had grave doubts about government wildlife agencies. To begin with, Leo's feelings about the U.S. Fish and Wildlife Service were colored by the skepticism most North Country Yankees have about federal agencies of any sort. As one North Country newspaper editor put it, the general feeling is that if a car with federal plates pulls into your driveway, the best thing to do is turn off the lights and slip out the back door, because you're probably better off without the federal government's particular brand of help.

Leo's own feelings about government wildlife management were further tainted by his experiences with New Hampshire's state agencies, which had been mixed at best. His encounters with wildlife law enforcement officers were only part of the story. Leo had spent countless evenings and weekends over the previous fifteen years trying to improve waterfowl habitat on the lake, and he felt that these efforts were underappreciated by the state wildlife agencies. He had started doing the work in the early 1970s, when a friend named Armand Riendeau convinced him to join a small hunting club called the Umbagog Waterfowlers Association. The UWA's main activity, other than talking about hunting, was planting wild rice in Umbagog's shallows to improve the waterfowl habitat. Riendeau had started planting rice in the 1960s despite assurances by state wildlife officials that he was wasting his time. The rice took and spread, and the club did too. Through the late 1970s and 1980s, Leo, Don, and their fellow club members spent much of their spare time sowing rice, building duck nesting boxes, and organizing game feasts to raise money to buy the rice and boxes.

Leo is proud of this conservation work, prouder still of his son Keith's decision to go to college to become a fish and game conservation officer. But he's bitter that wildlife officials have shown so little interest in the UWA's efforts, despite that the rice has proven a fruitful food source for the ducks and that most of the boxes host nesting pairs of ducks each year. He says, "Fish and Game don't want to take the time to see our rice and box work or hear our duck numbers. But they're real quick to ask for our licenses and check our bag limits." It also irks him that state agencies have repeatedly denied the UWA's request for a different duck hunting season in the northern part of the state. He says the present, split season – a week in mid-October and a week in mid- to late November – is designed mainly to coincide with migration patterns in the state's warmer coastal and southern lake regions. This forces the Umbagog hunters to hunt either too early, when the mainland migration is not yet coming through, or too late, when, in Leo's words, "the only ducks on the lake are the ones wearing ice skates." This has peeved Leo and his fellow UWA members no end.

"We raise 'em up here," says Leo, "and they shoot them down there."

After dark has fallen and Rene, Billy, and Marcel have left, Leo sits on the couch talking. Don is cleaning his rifle on the kitchen counter. Along with the rowdy camaraderie of the earlier story-telling, gone too is some of Leo's ire.

"Sometimes I think I should just quit the ducks," says Leo. "At first I thought, You gonna tell us we can't hunt Milan Meadows or Sweat Meadows or Leonard Pond because now it's your refuge? To hell with that. But with the ducks the way they are, it can't hurt. We used to have so many. Now there's enough for about two days of hunting, then they're all killed or gone away. Sometimes I think they should just shut it down for five or ten years, let things re-establish themselves, then open it up again. Maybe it'd be like it used to. I mean, I know it won't ever be the way it used to. The way the ducks would come down in the hundreds. It will never be the same. What my father had, we will never see. What we had, my son will never see. But if the birds got a break, maybe they'd get back the way it was a few years ago.

"The problem is, once they stopped the hunting, they wouldn't ever open it again. The Sierra Club and all the environmentalists would get into it, and we'd never get back in there. They'd be like, 'Oh, my pet ducks!' People from away come up here and see the birds, and it's like they think they're theirs. They're pets, and if you shoot them you're murderers."

He shakes his head. "It's like with the eagles. On Leonard Pond now they got zones and buoys – all kinds of regulations. You get caught harassing them or going by too close, too fast, you get in a lot of trouble." He shakes his head again. "The eagles. It's like the Shrine of Fatima.

"I mean, people, come up, enjoy the nature. But don't try to tell us what to do. This is not a park. We were raised up here. This is our country. We're not gonna stop doing something just because you say so. Not when we've always done it."

2

Working for the Company

Like his father and both grandfathers, Leo Roberge's first steady job was with the Brown Company, which had managed the pulp and paper mill complex at Berlin since the 1880s. Unlike other men of his generation, however, who were lining up for manufacturing jobs in the mill, Leo landed a job in the woods. It felt right to him. His mother's father had muscled a crosscut saw for Brown in the 1920s, and he liked the thought of following in those footsteps.

This was a lucky break. Paper company logging jobs were rare by the mid-1970s, because loggers were expensive to insure and the equipment was increasingly costly. As were other large timber firms, the Brown Company was finding it cheaper to contract with independent loggers to cut its own land, and to buy much of the wood needed for its mills on the open market. All counted – foresters, loggers, scalers, and truck drivers – Brown had maybe one hundred employees working in the woods in western Maine and northern New Hampshire in the 1970s, down from more than one thousand fifty years earlier. Leo's chance came in 1975, when the company decided to experiment with whole-tree harvesting and chipping, a technology that was just then emerging in the Northern Forest. The woodlands division purchased a fleet of sophisticated equipment and hired a seven-man crew. Leo, strong and fit after a hitch with the Marines and four years of pouring concrete, came on as the "ground man."

Brown's whole-tree system was entirely mechanized; except for the ground man, the workers rarely left the enclosed cabs of their massive hydraulic machines. The bulk of the work was done with

15

a powerful "feller-buncher," so named because it could cut and pile whole trees in one step. The feller-buncher, which rode on tank-like crawler tracks, resembled a large backhoe with a large hydraulic arm. Instead of a shovel, the arm was tipped with a grapple claw positioned over a set of giant shears. The operator would drive the feller-buncher up to a tree, seize the tree with the grapple, shear it with the giant pincers, and lay it down in a pile. The whole sequence took about ten seconds. Another crew member would follow through the woods in a grapple skidder, a heavy-duty rubber-tired woods tractor hinged in the middle for moving through tight spaces. The skidder had only a small blade, for it did most of its work with a large grapple claw that hung off its back. The operator would back up to a pile of downed trees, grab them with the claw, and drag the load to a clearing where the huge portable chipper sat. There the chipper operator picked up the trees with a mechanical arm and fed them, branches and all, into the chipper's grinding maw. A steady barrage of one-inch chips shot out a chute on the chipper's back end and into a tractor trailer to be hauled to the mill.

When he first joined the crew, Leo made four dollars an hour cutting large branches off downed trees with a chainsaw and running a skidder when needed. Slowly he moved up in rank, learning to run the feller-buncher so he could fill in when the regular operator was out. On those days he earned two dollars an hour more than when he was working on the ground.

"We were experimental, and were more or less our own bosses," Leo says. "We were separate from the other guys. We got paid by the hour, not the cord like they did. So they didn't like us too much. They thought we were pets, though they probably made more money than we did. But when we needed stuff we got it. If something broke down, a mechanic came out and worked on it. If you're on a regular gang and the skidder breaks down, you fix it yourself.

"We got good at it. We'd average about a hundred-fifty cords a day. Once we cut two hundred and ten. The machines never stopped working. We'd stagger our lunch breaks and always have a spare operator, so we could run the machines ten-and-a-half hours a day."

A cord of wood is a pile of logs four feet wide, four feet high, and eight feet long. In an average day, the crew could cut four acres of forest and fill a dozen tractor-trailers with wood chips. In 1979, the Brown Company started a night crew, and Leo won a full-time job running a feller-buncher on the graveyard shift. All night long he maneuvered the awesome machine through the woods, shearing trees in the glare of its spotlights. Despite the mechanization, it was rough work. Brown would send out a mechanic for big problems, but the hydraulics needed constant tinkering. Leo had to adjust valves, fix hoses, and change the crawler tracks that got damaged by clattering over the rocky terrain – grueling, tense work, particularly in winter. More than once Leo found himself crawling out on the machine's massive, icy arm with a wrench and grease gun.

After a long winter working nights, Leo got his first-class operator's license, a two-dollar raise, and a spot on the day shift. Elaine, his wife, was also working, and the family had full insurance, benefits, and regular vacations. By 1985, Leo was building seniority.

"We were the cat's ass," he says. "Usually a crew has one guy who's a dubber, who's always dogging it. But not on that crew. Out of those seven original guys, in ten years I don't think one guy missed more than five days of work. I missed only one, and that was because my ride was hung over after a union party. It was a good job. We knew what we were doing. We were like the 'A' Team."

When Leo's grandfather was sawing spruce for William Robinson Brown in the 1920s, the Brown Company owned 3.75 million acres of forestland in northern New England and Canada – an area larger than the state of Connecticut. This, however, was no arbitrary figure; it was the precise amount of land Brown needed to supply the company's pulp and paper mills in New Hampshire and in La

Tuque, Quebec, without ever cutting the forest faster than it was growing. This concept of "sustained yield" was relatively new. During the first timber boom in the Northern Forest in the nineteenth century, loggers had simply cut all the best trees and then moved on. But Brown took pride in being a pioneer of the emerging science of forest management. He was the first timber industry executive to hire a professional forester, and he recruited all his top staff from Yale's renowned School of Forestry. Brown's foresters performed original research on tree growth and genetics, ran the nation's largest private tree nursery, and wrote papers with titles such as "Scientific Management in Lumbering," "Making and Driving Long Logs," "Making and Driving Pulp Logs," and "Bark and Its Uses." For several years the company sponsored annual conferences that drew foresters and researchers from around the world.

The foresters' primary goal, however, was to provide a steady stream of wood from forest to mill. To supply the mills in Berlin, the Brown Company employed hundreds of loggers working out of forty camps scattered throughout northern New Hampshire and western Maine. W.R. Brown organized the region's first woods safety program, built efficient camps, hired professionals cooks (including Leo's grandmother), and provided his men with the best equipment. Nevertheless, the work was brutal, largely because most logging was done in the winter, when it was easier to slide logs over the frozen ground.

Every morning at first light, Leo's grandfather and his fellow woodcutters trooped along rough haul roads and into the woods. Teaming up on two-man crosscut saws called "misery whips," they would fell spruce trees, chop off the branches with axes, and saw the trunks into logs. A teamster followed with a horse to drag the logs out to the haul roads and onto a large sled. A team of horses or a steam-powered tractor then skidded the loaded sled along the road until they reached the banks of one of the rivers – the Magalloway, Rapid, Diamond, or the Dead – that fed into Lake Umbagog. When the spring thaw arrived, the loggers rolled the logs into the rushing water, which carried them downstream and into the lake. Once on the lake, Brown Company steamer ships corralled the logs together with long booms spread out across the water and towed them down to the dam at Errol. When the dam operator opened the sluice gates, the rush of water drove the logs down the Androscoggin River to Berlin. From landing to mill, the log drives were accompanied by river drivers wearing spiked boots and carrying long poles to free the jams. Wood from more accessible areas reached the city by railroad.

During the first quarter of the twentieth century, the Brown Company's mill complex in Berlin and neighboring Gorham – first built in 1867 and purchased by William Wentworth Brown in 1888 – had become one of the largest in the world. It produced an astonishing variety of paper goods, including newsprint, brown kraft paper for wrapping and making bags, sandpaper, wax paper, stationery, and, later, the ubiquitous Nibroc folded paper towels that are still used in restaurant and service-station bathrooms across the country. A lumber mill sawed long logs into shingles, clapboards, window frames, sashes, and doors. Smaller plants pounded pulp into flour and molded it into conduit pipes, twisted wood fiber into twine, built paper-towel dispensers, and mixed chemicals. Power plants burned scraps and sawdust to provide electricity, while repair and salvage shops fixed equipment and recycled materials. Pulp was not only used on site, but sold to other paper mills around the country. The company owned a railroad and five hydropower dams, and its research facility lured engineers from all around the globe. Sales offices marketed Brown products from Boston, New York, Atlanta, St. Louis, London, and Paris. In all, the mills at Berlin employed some five thousand men and women, making it New Hampshire's largest employer. The company generously funded sports, arts, and education programs in the city of Berlin, which had twenty thousand inhabitants.

When the Depression hit, paper sales plunged. But the Browns had developed a loyalty to the people of the North Country, and they kept the mill running so their neighbors could keep working. As orders slowed, railroad cars full of paper rotted on the rails. In 1934, the Brown Company filed for bankruptcy. To prevent the company from closing, the state legislature convinced Coos County to guarantee an emergency loan. With this borrowed time, Brown started looking for a buyer.

Through the Depression and World War Two, the company changed hands three times. With each transaction, the Brown name and local management remained intact. But the successive owners (which for a time included an Italian conglomerate) had little interest in Berlin or, for that matter, in making paper. To them, Brown Company was an investment to be kept just long enough to wring out a profit. Unaided by capital improvements but required nonetheless to show a profit, company managers were forced to sell land, cut employees, downsize the research and woodlands divisions, and close the sawmill and commodity factories. In the 1950s, the company sold its mill and timberlands in Quebec.

By the time Gulf & Western bought the Brown Company in 1967, the firm was a shadow of its former self. It employed only sixteen hundred people, a 65 percent decline since the war. Most of the land had been sold to finance earlier buyouts, and the aging mill was dirty and slow. The Clean Air and Clean Water Acts of the early 1970s required the installation of air and water scrubbers costing tens of millions of dollars. Gulf & Western made only the improvements demanded by law, while cutting expenses wherever possible. By the late 1970s, Gulf & Western, no longer interested in carrying what had become one of the Northern Forest's most problematic mills, starting shopping for a buyer.

In 1969, two men named Brent Halsey and Robert Williams founded a new paper company in Richmond, Virginia. James River Corporation differed somewhat from other paper companies. For one thing, the firm did not like to own large tracts of land, preferring to buy most of its wood on the open market. For another, part of the company's strategy was to buy old, slow paper mills. Halsey and Williams would purchase such mills relatively cheaply, figure out what the facilities could do without any major upgrades, and convert them to produce high volumes of three or four specialty products: computer paper, industrial films, and absorbent products such as paper towels and toilet paper. James River made its first foray into New England in the 1970s, when it bought two paper mills in Old Town and Jay, Maine. At the time, James River was one of several large corporations that were buying smaller timber companies in the Northern Forest.

When James River officials looked at the Brown Company's Berlin complex in the late 1970s, they liked what they saw. The mill was surrounded by working timberland, it was in a city with a reliable labor pool, and the old machines could still produce long runs of low-grade papers. And so in 1980 the Brown Company mills and land changed hands for the fifth time in as many decades. James River immediately sold approximately a half-million acres in Maine and New Hampshire to the paper company Boise-Cascade to help pay for the purchase, reducing Brown's once sweeping empire to fewer than 200,000 acres. The Brown Company name was formally dropped, and the "JR" logo appeared on the yellow brick office buildings along Berlin's Main Street. For the first time in a century, no Browns were connected to the mill.

For a while the Brown family's paternalistic culture remained intact, partly because James River retained mill manager Edgar Dean, who had run the complex for several years. Dean convinced James River to continue to support youth hockey, arts, and other civic programs, and to honor a lingering commitment to help fund a new bridge over the Androscoggin River. Then in 1985, James River replaced Dean with a Californian named John Shank.

Jeffrey Taylor, who was Berlin city planner at the time, was one of several municipal officials whom Shank invited to lunch shortly after arriving in town. Taylor recalls that Shank opened the discussion by reminding the officials that a high cyclone fence surrounded the millyard. "You know that fence?" Shank asked. "You gentlemen take care of everything outside that fence and I'll take care of everything inside it."

Taylor says it was impossible to overstate the symbolism of the fence. "When Shank arrived, there was a completely different corporate philosophy," Taylor says. "The Brown family traditions were gone. James River still did some community things, but the level was dramatically decreased. From the standpoint of the profitability of the mill, it might have been a good thing. But for the people, it was a real disappointment." A long-time senior manager who had retired just before the James River purchase remembers an uneasy sense among Berlin's inhabitants that Shank "just didn't give a damn." Shank wasn't in Berlin to play the part of doting uncle; he was there to make a profit from an obsolete mill and 200,000 acres of land. He continued the layoffs and further narrowed production to a few profitable lines.

Before long, the new management team noticed Leo's whole-tree logging team. The crew's production had been dropping as they had to go farther and farther to find good stands to cut. The aging equipment required regular repairs, and James River was buying more and more wood from independent loggers. But there was an even more direct problem. To run the mill efficiently, the manager demanded consistent and clean pulpwood. Chips processed in the forest included debris and dirt that could be removed only with expensive cleaning and bleaching. A few months after Shank came to Berlin, in 1985, the company sold the whole-tree harvesting equipment and disbanded Leo's crew.

"They went from shaking your hands to cutting your throat," Leo says. "Shank came in with an ax, and he started chopping. The man was gross. At the time, I was making twelve-fifty an hour in the woods, a good living. I had the weekends off, got to see my kids, I could travel, there weren't these strange shifts. You get something you like and then you lose it."

Unable to find a position on the company's one remaining logging crew, Leo went to the mill and was assigned to the "salvage pool," a group of on-call laborers who rotated through different areas of the plant. Paper mill jobs were the highest paying blue-collar jobs in northern New Hampshire, but now Leo was under a different union contract. He lost all his seniority and took a sharp pay cut. Some weeks he knew he would be working. Others he'd get a call in the afternoon to come in for the graveyard shift. He moved from station to station as production demanded, running paper machines, driving a forklift, working in the boiler rooms. In winter, he spent days at a time chipping ice from the grates in the company's hydropower dams so the river water could flow through the turbines. Sometimes he had to clean up rooms where lime was stored, trying in vain not to perspire in the sweltering heat, because lime sticks to sweaty skin and burns. Some shifts were twelve hours, but at least he kept his job. Many of his co-workers were not so lucky.

"Cutting people made them look good," Leo says. "They told us that was the way to keep the rest of us working."

Looking good became especially important to James River in 1990, when the company decided to sell the mills in Berlin and Gorham. By then, James River Corporation had ballooned into a Fortune 500 giant with factories in two dozen states and a dozen foreign countries. Globally, pulp prices and production had climbed steadily from 1980 to 1988, but by 1990 a paper glut was causing an industry-wide recession. Despite substantial investment and some modernization, the Berlin complex was still losing money. James River was undergoing a major restructuring program, and Berlin no longer figured in the corporation's strategic plans.

By cutting costs, James River officials hoped to show prospective buyers a good cash flow so the Berlin mill would sell quickly. They hoped to pass the remaining upgrades on to someone else. The paper machines were still too slow and too small to compete with increasingly sophisticated mills in Maine, the southeastern United States, Canada, and overseas. The Federal Energy Regulatory Commission's license for the company's five hydropower dams was coming up for renewal. There were chronic tensions between management and labor. Most important, the pulp mill's massive recovery boilers, pressure cookers that recycled the chemicals used to dissolve wood chips, required extensive improvements to satisfy air pollution laws. Upgrading the boilers alone would cost more than $70 million. The entire industry knew this, and no buyers came forward.

Although financial difficulties and owner turnover had been part of the mill's story since the Depression, residents were particularly uneasy this time around. Diamond International had recently been dismantled and its lands sold, the nation was slipping into recession, and half the jobs in the county depended either directly or indirectly on the Berlin mill. If a huge company like James River couldn't make the plant profitable, who could? Rumors ranged from a Japanese buyout to an outright shutdown.

James River did manage to sell its other New Hampshire holding, the much smaller Groveton mill thirty miles to the west, to Wausau Paper Company of Wisconsin. After eighteen months on the market, however, the Berlin-Gorham complex had attracted no offers. In 1991 the state, afraid of losing its fifth largest employer, passed legislation offering to back a $25 million low-interest loan if James River would take the plant off the market. The company never took advantage of this offer, but it did patch up one of the boilers, shut down the other, and stopped talking so publicly about selling the facility. Layoffs continued until, by 1993, the company had fewer than thirteen hundred employees in Berlin and the city's population had shrunk to ten thousand.

"Shank's gone now," says Leo. "He did his job. He's out in Oregon or something, and I'm in the mill doing shift work. Quite a change. Climb up the ladder, you get kicked right back down again. Now there's guys with seventeen years who are hardly getting any work. But life goes on. There's not much you can do about it.

"My son Keith, I told him, 'This mill is not a place for you.' When the kids are out of school, maybe I can do something else, get out in the woods again, maybe operating some machinery. I'd like to try something on my own."

3

Fragile Alliances

As James River's commitment to Berlin and Gorham wavered in the late 1980s, many people grew increasingly uneasy about the future of Lake Umbagog. The cash-strapped company's land around the pristine lake, where single lots were valued at up to $50,000, was its most valuable liquid asset in New Hampshire. The rest of the state was experiencing a monumental growth boom, and Umbagog's distance from population centers no longer seemed so great. Local residents worried about vacation home buyers cutting off access, and environmentalists and wildlife experts feared losing the state's most biologically diverse inland ecosystem.

It wasn't the first time the lake had been threatened. There was a proposal in the 1960s to mine the lake's shallow bottom for its diatomaceous earth, a fine silt of algal remains that is used for industrial filters and polishes. In the 1970s, a group of investors proposed building a floating Japanese restaurant. And earlier in the 1980s, Union Water Power Company, which owns the Errol dam, wanted to run a 17-foot-diameter water pipe from a dam on neighboring Richardson Lake to Umbagog's north shore, install a hydro plant at the base of the pipe, and string high-tension lines along the north shore. Each time, area residents joined with environmental groups from downstate to fight off the development schemes. These alliances were uneasy, but the influence of the professional environmentalists combined effectively with the determination and common sense of the locals. The earlier successes set an important, if fragile, precedent. They also strengthened the locals' conviction that the lake was really theirs to defend and enjoy.

26

In the late 1980s, however, two events – one huge and nebulous, the other tiny and tangible – broke the pattern. As New England's real estate market peaked, developers, looking ever further north for "wilderness lots" to sell to newly prosperous vacationers, began to eye Umbagog. From the locals, this rather abstract economic threat drew the sort of response that rumors of a ghost might: They argued over whether it was real. Some, looking at the heavily developed shorelines of Lake Winnipesaukee to the south, Rangeley to the northeast, and even tiny Aker's Pond right in Errol, saw the development threat looming large. Others, looking at Umbagog's peaceful shoreline and remembering their previous victories and the long history of stable ownership, said such changes couldn't happen here. Meanwhile, conservation organizations and agencies, having been caught off guard by the sale of the former Diamond lands in 1988, felt the threats quite keenly. Thousands of acres of New Hampshire farm and forest land were being converted to house lots every year, and the conservationists knew that no magic circle protected Umbagog.

The other development came twig by twig. In 1988, a pair of bald eagles built a nest in a tall white pine tree on an island in the lagoon called Leonard Pond. It was the first time in forty years that bald eagles had nested in New Hampshire. When a first eaglet died, biologists replaced the dead bird with one hatched in a zoo. As the eagles successfully raised the adopted eaglet, the birds became an irresistible symbol of what a wild, unique, and valuable habitat Umbagog was. The first pair of eagles, along with a second pair that began frequenting the southern end of the lake in 1990, soon became a pivotal point in the argument that environmental groups were making to large landowners: that Umbagog should become a wildlife refuge. Biologists had long been hoping to establish a refuge there, primarily to protect waterfowl. Development pressures and the presence of the eagles provided a new sense of urgency.

Both Maine and New Hampshire had recently created public land acquisition programs. In 1988, staff from these programs –

the Trust for New Hampshire Lands and N.H. Land Conservation Investment Program, and the Land for Maine's Future Board – started to negotiate quietly with officials from the U.S. Fish and Wildlife Service, James River Corporation, and the two other large landowners around the lake, Seven Islands Land Company and Boise-Cascade. A larger Lake Umbagog Study Team, which included these interests as well as several biologists and local residents, started mapping out the refuge. James River officials emphasized that they wanted to protect the lake from development, but would not give up the right to manage for timber.

After many months, the team developed a plan to protect some 16,000 acres using a variety of land conservation techniques. About half the land, the critical wetlands in places such as Leonard Pond, would be bought outright by the U.S. Fish and Wildlife Service. A 500-foot buffer on the western shore would be protected by the state of New Hampshire. The state and federal agencies would also buy conservation easements on thousands of acres of upland forest. The easements would leave ownership in the timber companies' hands and allow them to harvest trees, but forbid development. It was an extraordinary and unprecedented plan, made possible largely because several James River officials – Brent Halsey, Jr., who was in charge of all the company's New England timberlands, Brad Wyman, woodlands manager in Berlin, and forester Phil Bryce, who worked directly under Wyman – understood the importance of Umbagog, and were willing to try something new. It was a model that many believed could be used elsewhere.

When the Fish and Wildlife Service announced the plan to the public in the fall of 1990, area residents were not thrilled. Many protested about the possible restrictions on hunting and boat speeds and the end of private camp leases on paper company land. There was also broad resentment about federal agencies and public land in general, as expressed by state senator Otto Oleson during a tense hearing in Berlin's city auditorium: "I'm very wary of the federal government acquiring anything," Oleson said. "I'd like to see ninety-nine percent of the determination about how the lake will be managed made by the local people in this room." Oleson was applauded for his testimony, as were several speakers who voiced similar sentiments. But near the end of the hearing, some of those same people clapped after a young biologist named Jeff Fair gently reminded the crowd that the lake wasn't owned by the people in the room; the timber companies owned the land, and the timber companies wanted to sell. Fair's candidness seemed to strike a chord.

"James River has other things on its mind right now than saving Umbagog," said one man, who was wearing a "JR" baseball cap. "If the feds don't buy all that lakefront from the company, someone else will. And they're not going to invite us all in for a chat about their plans."

"You've got to understand," said another, pointing at the federal officials. "Everybody here distrusts James River even more than they distrust you. You just have to get the people involved."

Project manager Dick Dyer did just that, meeting with small groups of people in their living rooms, hunting clubs, and town halls. He still found plenty of opposition, but many residents were coming to support the idea.

"I'd hate very much to see it where a few rich people owned most of the camps on the lake," said Errol gas-station owner Everett Eames. "That could happen if something isn't done along the lines of this conservation plan." Luke Cote, who owns a sporting-goods store in Errol, said that he had once dreamed of buying land at the lake and building a development. But by early 1991, he was supporting the refuge. "You could make yourself a few million on development," Cote said. "But then you'd have to move away from here because it would no longer be so beautiful. If we don't protect something up here, there won't be anything left."

In June 1991 the U.S. Fish and Wildlife Service released its Final Environmental Assessment, which recommended proceeding with the 16,000-acre proposal. The following fall, the government bought its first parcel of land for the Lake Umbagog National Wildlife Refuge.

Jeff Fair's Isuzu Trooper is parked on Success Pond Road at the southern end of Lake Umbagog. Fifty feet to either side of the truck is a sheet of plywood propped up with a stick, and stencilled in orange: NH FISH AND GAME HUNTER CHECK AHEAD. The truck's rear bumper sports a single sticker that says "Nature Bats Last." A Brittany spaniel is pacing in the front seat. Whenever the dog puts his feet up on the open window, which happens about every two minutes, Jeff barks, "Tripper! Kennel up." The little tan-and-white dog curls up in the passenger seat for a minute or two, then gets curious again.

Jeff is wearing wool: charcoal gray pants, vest, and hat, and a red shirt. He has wavy, longish dark brown hair and a beard, and wire-rimmed glasses. A blue bandana is tied around his neck, and rubber-soled leather boots cover his feet. On the frozen mud next to the truck is a battered three-burner camp stove with a black speckled enamel coffee pot. We hold plastic cups in our gloved hands, hoarding the coffee's steamy heat. The lake, barely visible through the spires of spruce, sends an icy wind. It's November and feels like it.

From where we're standing, Success Pond Road runs south to Berlin, parallel with the eastern bank of the Androscoggin River, and through the unorganized townships of Cambridge and Success. In dry seasons, it's a dangerous place to drive. Loaded log trucks rumble down the middle of the gravel road as if they own it – which in a sense they do, since the road and all the land between here and Berlin is managed for pulp and sawlogs by James River Corporation. Come hunting season, however, pickups outnumber log trucks. Locals know that this road provides the most direct route from Berlin to the lake. Its innumerable feeder trails and skidder paths reach into the empty backcountry, where poplar and maple regenerating in patch and strip cuts draw in hungry moose and deer. When the deer hunting parties come through in the morning, Jeff gives each one a number and asks them to watch for moose, as well as for bear, grouse, and pheasant. On the way out in the evening, they'll stop and report what they've seen. Jeff will tally up the sightings and give them to the N.H. Fish and Game Department, which uses the data to monitor wildlife populations and hunting pressure.

Jeff has spent a lot of time counting animals, mostly during a ten-year stint as director of the Loon Preservation Committee. With its upscale board of directors and Audubon Society sponsorship, the loon committee is usually associated with Lakes Winnipesaukee and Squam, a hundred miles south. But it was at Umbagog that Jeff did his best work, slowly returning the lake to its status as the most productive loon breeding ground in New Hampshire. Now freelancing as a consultant and writer, he spends five days a year staffing a Fish and Game biological check station. He always asks for a spot near Umbagog.

A red Ford pickup truck pulls up and Jeff steps into the road. Behind the wheel is a tall man in his late thirties, with a light mustache and bloody hands. In the bed of the pickup are an all-terrain vehicle and a big male white-tailed deer, its antlered head hanging off the side. The eyes stare; impossibly bright blood drips from the mouth. A single dark bullet hole in the animal's side is caked with blood and matted hair.

"Wow. Nice buck," Jeff says. "How big you figure?"

"I don't know, maybe two hundred," the driver answers. "He's probably three or four years old. He was back in a ways. Had to use the four-wheeler to get him out."

Jeff pulls back the animal's blood-crusted lips. He counts the teeth, then probes the thick neck as if giving a massage.

"Age sounds about right. His neck isn't swollen, though, so he wasn't in the rut."

The driver smiles. "Stop by the house later, maybe you can talk my wife out of the liver." He puts the truck in gear and drives off, the dead deer bouncing slightly as the truck rattles over the washboards in the road.

The successful hunter, Jeff explains, was Paul Thibodeau, operator of Union Water Power Company's Errol dam. Jeff got to know him when he was working for the loon committee. They enjoyed each other's company, and started hunting together once or twice a year. Now that Union is one of Jeff's consulting clients – he's studying the impact of dams on water-dependent wildlife – Jeff sometimes stays with Paul and his family at their home next to the dam. Jeff shares a few anecdotes about the large man's gentle manners and his unassuming love for the North Country.

It seems incongruous, this career environmentalist looking forward to eating fresh deer liver with a friend whose job it is to raise and lower lake levels – a process that can threaten wildlife in general, and the loon in particular. Yet when it is mentioned, Jeff bristles.

"Look. Locals like Paul have goals much more similar to the environmentalists than people think. They might have a different idea on how to achieve them, but they're basically the same: to keep it like it is. The locals care just as much about wildlife and conservation as folks from Audubon. They just don't want to be preached to."

When Jeff Fair started working for the loon committee in 1980, the Umbagog birds were in trouble. Loons, which move clumsily on land and easily in water, nest in low areas along the shoreline so they can slip in to fish. When Union Water Power closed the gates at the Errol dam and the lake rose too quickly, the loon nests would be drowned; when the gates were opened and the lake fell, the distance from nest to shore left the birds vulnerable to predators. The changing levels also affected ground-nesting ducks. Ringed-neck and black ducks were flooded out of their nests, and the wild rice planted in the rivers and shallows by Leo Roberge and the Umbagog Waterfowlers Association was washed out.

Communications were strained between the power company and local camp owners and sportsmen. Leo recalls that a Union official once asked him to explain why a duckling killed by drowning in the egg was any worse than a mature bird killed two years later by a hunter's shotgun. The question revealed a gap between the two men that Leo simply couldn't understand. But Jeff knew how to talk to the company, and he had the political weight of the Audubon Society of New Hampshire behind him. With support from the hunters, the Loon Preservation Committee helped negotiate an agreement with Union to keep the lake from rising more than 6 inches or falling more than 12 inches, whenever possible, in any twenty-eight-day period. A handshake deal at first, the reference to consistent water levels was later included in Union's license with the Federal Energy Regulatory Commission. Jeff also started launching loon rafts (floating docks piled with marsh grasses to simulate natural nesting sites), posting notices about nesting areas, and talking with local people about the birds' need for privacy.

He also began hunting again, something he hadn't done since high school. Hunkering down in blinds with Armand Riendeau and his friends, he listened to them talk about the land and the lake. Their skills of observation and intimate knowledge of the backcountry impressed him, as did their unease as Umbagog was increasingly studied and monitored by downstate biologists and wildlife officials. But Jeff knew the downstate perspective, too. His wife is a Forest Service biologist in the White Mountain National Forest, and he has spent most of his professional life working closely with conservation groups and agencies. He was equally sure of the environmentalists' sincerity and motives.

"The groups get teased apart," he says. "The hunters tell Fish and Game and Audubon folks what they see out in the woods, but the biologists are trained to be more methodical. So their attitude has been basically, 'Well, thanks, but we don't quite believe you, because we haven't inventoried that area yet.' We cross paths a lot up here: biologists, hunters, bird watchers, loggers, camp owners. But people don't think about sharing their cards with each other, so it ends up in factions. It's all like a poker game. But it doesn't have to be."

The wildlife refuge, Jeff says, raised the stakes. As a key member of the Lake Umbagog Study Team, he was convinced that some measures beyond outright federal acquisition would be necessary. This was not only because of the economic need to keep some of the land available for timber harvesting, but also because of the prospect that heavy federal involvement would upset the delicate triumvirate between locals, state conservation groups, and the landscape.

"There are plenty of examples of places that have been 'saved' by federal intervention," Jeff says. "When the feds got involved with the refuge plan, there was this one guy from the U.S. Fish and Wildlife Service who started out convinced that the duck hunters and camp owners would be dead set against the refuge. He simply wouldn't accept that this wasn't just a bunch of shoot-em-up guys out there drinking and shooting ducks. He said it would be us against them all the way. He figured that the people up here couldn't appreciate the threats to the lake. But very few people up here haven't gone far enough south to see what can happen. They've seen the signs down on Squam that say 'No Hunting' and 'No Trespassing.' They see development coming. They just wish they didn't have to worry about it. But that doesn't mean they don't know what's up."

Tripper is growing increasingly restless, sticking his nose out of the open window and whining. Jeff opens the door and the dog springs out of the truck, trailing a length of neon red surveying tape from his collar; you're never so far north that a dog can't be mistaken for a deer, Jeff says wryly. Down the road, the dog stops suddenly near a pile of slash. We admire his point – Jeff has been training him to hunt since the dog was weaned – and walk up the road to see what he has found. It turns out to be a partially decomposed grouse. Jeff whistles and follows the dog back to the truck. He opens the passenger door and the dog jumps in, turns around twice, and lies down with a grunt. Jeff pulls his glove down to look at his watch. It's after five and nearly dark, but two of the day's hunting parties have not yet returned. He returns to the coffee pot and refills. Reaching into the cooler through the truck's open back window, he digs out a silver flask and splashes a half shot of Canadian Club into the coffee. He turns up his collar and settles back against the vehicle's front fender.

A place like Umbagog, Jeff says, is easy to love, easy to understand. It takes a little longer with the people. When he first came here as a graduate student in 1973, the locals were friendly but somewhat distant. When he returned in 1978 after a few years out west, the atmosphere was different. Jeff recalls being in the Errol diner and running into Warren Jenkins, the local conservation officer, who had always seemed somewhat remote. But at the diner that day, Jeff says, "Warren shook my hand warmly and said, 'We heard you were coming back around. Good to see you.' He talked for a minute or two and returned to his table. No big deal. But it meant a lot to me. I guess I had to come back before Warren would really respect my love for the lake. He could see that it was more than just a college project. Because I came back.

"Once you get to know these people, they'll do just about anything for you. I used to go duck hunting with Armand in the morning, and then he'd help me pull loon rafts off the lake in the afternoon. When he ran his car off Route 16, I went to the memorial service. Guys kept coming up to me and saying, 'Hey, if you need help pulling them loon rafts, just give me a call.' Too often we don't allow for that kind of crossover."

4

Birding

The Androscoggin is running high this morning, swollen with two days of May rain. We put ashore at the mouth of a small, tea-colored stream and pull the canoe up a mud bank and into the spruce forest. Then we shoulder our packs, drape binoculars around our necks, and walk into the understory, looking for the yellow tags that will mark our course. It's five in the morning, cloudy and dark. We nevertheless soon find a yellow ribbon tied shoulder-height around a large spruce.

The first call we hear is easily identified: the round, eerie hoot of a loon crying from the river fifty yards away. The next one is tougher: a quick rising song, wheeta wheeta WHEET. The bird sings some thirty yards away, hidden in thick stands of spruce and fir. Carol Foss, Audubon Society biologist and Ph.D. candidate at the University of Maine, lover of Lake Umbagog and of wet soft-wood forests, whispers, "A magnolia or a yellow." She bends her head and listens. The bird calls again.

"Magnolia," she says. She writes in her notebook a four-letter code, MGWA, to note that she has identified a magnolia warbler. Her research assistant, Krishnan Sudharsan, lifts his clipboard, flips open the aluminum cover, and pencils a dot onto the lower-left-hand corner of a piece of graph paper to represent the bird's location in this plot of land. Over the next three minutes Carol identifies and records the thin high song of a golden-crowned kinglet, the rising rattle of a northern parula warbler, the flutelike song of a Swainson's thrush, and the short complaint of a yellow-bellied flycatcher. Six weeks ago, these woods were cased in snow,

36

silent but for the chickadees' chirp and the light smacking and chipping of northern juncos. Now the woods are crowded with warblers, flycatchers, vireos, thrushes, and other birds newly arrived from their 5,000-mile journeys from the tropics. From wintering grounds all over Mexico and Central and South America they come, crossing the Gulf of Mexico in April or May in night-long flights. After resting on the Gulf beaches, they make their way up the eastern seaboard to find breeding habitat. The males call eagerly, prolifically, as they stake out territory and sing and dance to attract mates. Those birds that come this far north have only ten or twelve weeks to establish territories, find mates, breed, and fledge their offspring.

During that time, Carol and Krishnan and two other crews will do these "spot counts" four days a week. Each crew covers one of twelve plots each day, walking up and down nine rows of nine spotting stations laid out at 50-meter intervals over a 50-acre plot. This damp spruce lowland along the Androscoggin is one of four lowland softwood plots that Foss is studying; there are also eight hardwood sites in the Nash Stream watershed 25 miles west, four mature and four that were clearcut within the last decade. At each station the field workers stop, identify every bird they hear or see, and note each species and its location. The method is rigorous, even exhausting. The terrain is rough, the weather unpredictable, the mornings long. In two weeks the black flies will be thick. Today, on an unseasonably cool day in the last week of May, the soggy ground chills our feet even through knee-high rubber boots and wool socks.

Along with a willingness to rise at three-thirty every morning, this regimen also requires some prodigious birding skills. In a typical morning Carol might hear a hundred birds, but she will see only three or four. Fortunately she knows bird songs as well as almost anyone in northern New England. She pegs many of them on the first phrase. She walks, listens, records, walks again. She and Krishnan thus generate each day a map of that day's sightings.

They note any nesting activity they see, such as adult birds carrying nesting materials or food for young or, if they are really lucky, fledglings. Later they will transpose the day's locations of each species onto a single, cumulative map for that species. As the weeks go by, the sightings for any given species will gather in clusters on that species' map, indicating the territories and nesting locations of each bird and mating pair. It is, says Carol, "just incredibly exciting" to see these territorial and nesting maps emerge.

We walk on a mat of shining club moss, a bright green plant that covers the ground like a carpet, padding our footsteps. Tiny rills, small streams, and wide muddy slews cross the terrain at random, emerging from beneath vegetation to trickle or slide along a few yards before disappearing again. In the sandy deltas formed by the streams grow fiddlehead ferns, which are just beginning to unfurl their curled heads; they look like hairy green bass clefs. Clusters of false hellebore burst from the putty-colored alluvial flats in fountains of wide, slick green blades. Every stump, every fallen log is covered with mosses, fungi, and lichens. We pass a hip-high stump completely covered, as if spray-painted, with a bright green growth so fine and thin that the comblike contours of the stump and the texture of its bark show through; when one of us breaks off a knob, it comes loose with hardly a sound, revealing soft blond wood. Logs litter the ground, forcing us to step constantly over or upon them. Only about half of these hold us; our feet plunge through the others, pulverizing the rotting wood into mulch. Trees grow atop fallen trees, sending their roots through them to the soil below.

All this fecundity is ironic, for the glacial soils here, heavy in granite, are actually shallow and acidic, inhospitable to most forms of life. The soil's thinness gives tree roots poor purchase (one reason why so many spruce and fir trees have fallen over), and their acidity keeps out most hardwoods and a wide variety of other plants. But what life these conditions do favor does well indeed. Where dead trees have fallen and let in daylight, young balsam fir and spruce grow profusely. The young knee- to waist-high trees grow in patches so thick you can't see the ground through them. The taller saplings pose a serious obstacle to forward progress. "Goggle time," Carol says when we enter those new stands, and we all slip our goggles down and push through, using our forearms and elbows to spread the slender poles apart as the stiff branches rake our faces. Our jackets and pants and caps are soon soaked.

At 5:47 we stop just below a low, rounded ridge about 3 to 5 feet high and perhaps twice that wide. It is one of several natural formations that run through these woods and then disappear, as if someone had started to build a series of poorly organized dikes and lost interest. From their relatively dry and deep soil grow hardwoods (maple, birch, a few red oak), white cedar, and on this one, a tremendous white pine. The pine's roots straddle the bermlike ridge. We stand in the middle of a small flat stream next to the tree and listen. We hear the deep, hesitant thump of a grouse beating his wings. The thumps begin at intervals of a second, then accelerate into a drum roll that fades away. Carol points along the line of the ridge, whispers "thirty meters," and Krishnan marks the corresponding position on the plot map. We hear a brown creeper, a tiny bird that will nest only in the curved pocket trapped behind the peeling but still-attached bark of a standing dead tree – a fact discovered here at Lake Umbagog in the late 1800s. We hear another Swainson's thrush, a Blackburnian warbler, and the loud melodic trill of the male winter wren. It's rare to actually see a winter wren, for they are tiny, well camouflaged, and reclusive. However, we do see an evening grosbeak as it flies over; its size and butterscotch plumage give it away immediately. This is the first actual sighting of the day. The grouse beats his wings again.

At each of the next several stops we see only one or two birds. At 6:39, in the middle of a thicket of ten-foot spruce, amid poles so dense we stand separated as if by jail bars, Carol wonders aloud in a soft whisper whether the cool weather will make this a slow morning. Krishnan takes another temperature reading. It has dropped to thirty-eight. For two minutes we hear nothing but our own breathing. We move on. For over an hour we hear few birds. At 8:03, we stop briefly to eat some granola bars and drink some water. As we eat, the clouds break slightly, admitting some sun. The sunshine, though it remains sporadic, seems to revive the birds. On the next half-dozen stops we record yellow-rumped, Nashville, Canada, Blackburnian, magnolia, and black-throated blue warblers, white-throated sparrows, several water thrush, a hermit thrush, a purple finch, a grouse, a nuthatch, a yellow-bellied flycatcher, a northern dark-eyed junco, another grouse, and an osprey, which screeches overhead as it hunts the river. These are most of the species we'll identify today; altogether we'll count ninety-eight individuals and almost twenty species in six hours of walking. We won't hear or see a single bird Carol can't identify.

$\mathcal{B}$ack at Carol's cabin, we sit before bowls of soup and a pile of grilled cheese sandwiches. The cabin, once a hunting camp but now owned by the U.S. Fish and Wildlife Service and loaned to researchers, is exceedingly simple. The kitchen has an old refrigerator, old stove, old sink and linoleum counters, and the wooden picnic table at which we sit. The only other rooms are a tiny bathroom, two small bedrooms with bunk beds, and a screened porch facing the Magalloway river. Umbagog is a mile downstream.

Carol is famished. She says little for several minutes. She sips spoonfuls of steaming tomato soup and eats the first of three sandwiches. She takes off her hat, scratches her fingers through her brown hair, and puts her hat back on.

"I've been trying for ten years to figure out how to spend a whole summer up here. I love this place. It's an incredible array of habitat. There's the lake itself, the huge wetlands, and all the rivers coming together. You set that into this matrix of where the northern hardwoods mix with the coniferous boreal forest coming down from Canada, and it really becomes a magnet for wildlife."

Carol has been coming to Umbagog for over ten years in her work with the New Hampshire Audubon Society in Concord, but only sporadically and for short periods, and mainly to coordinate the work of volunteers and interns monitoring wildlife here. During the late 1980s and early 1990s, she worked to re-establish a population of ospreys and helped track the return of the bald eagles. In 1993 she took a leave of absence from Audubon to get a Ph.D. in wildlife ecology with Malcolm Hunter at the University of Maine at Orono. Hunter, a prominent conservation biologist in New England, had spent a lot of time evaluating the impact of timber harvesting practices on wildlife. Carol had become particularly interested in how songbird populations responded to different forest practices. She came to Umbagog to try to develop a method not just for counting birds, but for actually assessing their breeding success under different conditions and in pieces of forest with different logging histories.

Assessing breeding success of songbirds, she explains, is difficult to do both accurately and efficiently. For a long time researchers have used mainly nest searches or mist-net surveys. However, nest searches, in which researchers try to locate nests and then climb the trees to see how the nests are doing, require huge amounts of time for information on relatively few nests. Mist netting – catching birds in finely meshed nets in the canopy – also takes a lot of time, since the researcher must drape the nets in the right part of the specific kind of tree in which the species in question likes to breed, let the birds get caught in it, get the nets down to count the birds, and then try to untangle the birds from the nets without hurting them. Sometimes the birds get so badly tangled they can't be removed alive. "They're both good techniques for doing intensive work on a small piece of land," says Carol. "But in terms of applying them over large areas, there's no way. So I'm trying to come up with a less labor-intensive way of assessing reproductive success."

Getting up at three-thirty every morning to do six hours of goggle time, of course, does not exactly seem labor-unintensive. But if Carol's method of gauging breeding success works, it will contribute a vital tool, for the breeding status of songbirds has become a question of some weight in the Northern Forest. Songbird numbers have declined alarmingly over the last twenty years, and this decline has been blamed at least partly on the effects of forest fragmentation from development and increased logging in the northeastern United States. Signs of trouble were first noted in the mid-1970s, when some of the most common and numerous North American songbird species, including such colorful favorites of amateur birdwatchers as orioles and tanagers, began disappearing from urban and suburban areas of the Midwest and Northeast. By the mid-1980s, wide-ranging studies carried out by the U.S. Fish and Wildlife Service found that many species were declining throughout the United States, particularly east of the Mississippi. Then a researcher examining radar pictures of the Gulf of Mexico found that only half as many songbirds were making the annual migratory trip north in the late 1980s as had in the mid-1960s. The biggest declines were in "interior" species – those that tend to nest amid contiguous stretches of forest – such as many of the warblers and wrens. A smaller number of "edge" species, which prefer to nest where wooded areas meet clearings, increased. The declines were worst in the Northeast; in northern New England, over three-quarters of songbird species suffered significant population drops during the 1980s.

Scientists first explored three possible causes for these changes: the increasing fragmentation of the forest in the United States; the corresponding increase in edge habitat; and the deforestation of tropical wintering grounds. They quickly found that none of these factors could sufficiently explain the dramatic loss. For instance, researchers found that the drop in bird numbers far exceeded the reduction in U.S. forest cover, and that even pieces of forest that weren't shrinking or being fragmented were losing many songbird species. They also found that most of the interior species concerned will breed up to the forest edge, contradicting the notion that the increase in the ratio of edge to forest area left interior species with no places to nest. As to the losses in tropical forests, researchers found that the songbird species declining fastest were not always those that were losing their winter habitat. So the loss of forest cover here and in wintering grounds, though widespread, didn't seem to fully explain the declines.

With the most obvious possible factors eliminated, researchers began looking for some other, underlying problem affecting the declining birds. They found one: the brown-headed cowbird. A dweller of fields and other open spaces, the cowbird had earned its name by following cows to scavenge seeds from heaps of dung. Though it feeds mainly in fields, the cowbird regularly enters wooded areas to engage in "nest parasitism" – that is, it lays its eggs in other birds' nests, then flies away. The cowbird chick typically hatches sooner than the host species' chicks do, and with its head-start and a size advantage, outcompetes them for the food the parents bring. Female cowbirds regularly fly up to 400 yards into woods to find a nest to parasitize, and they have been observed flying as far as three miles into forest interiors to do so, making the interiors of forest patches up to several thousand acres vulnerable to parasitism. In patches smaller than that, cowbirds often lay eggs in the majority of songbird nests; one study of Illinois forests found cowbirds parasitizing high percentages of nests even in tracts as large as 5,000 acres. Only the center of very large pieces of forest seemed safe. Studies found that smaller tracts also made the nests more vulnerable to outright predation from such nest raiders as blue jays and raccoons. It seemed forest fragmentation was playing a role after all.

By chance, these findings reached a critical mass just as concern over the Northern Forest began to rise in the late 1980s. The decline of songbirds immediately became a weapon in the arsenal brought by environmentalists to the land-use debate. The Northern Forest has no spotted owl (for which most people, including most environmentalists, are glad), but the vanishing songbirds seemed to provide a direct, easy-to-understand example of how heavy cutting and widespread development could harm wildlife.

Carol Foss is counting birds partly to see if this connection is as clear as it seems.

Biology, she says, offers no simple answers. She expresses caution, for instance, about drawing too many broad implications from the studies that have documented the songbird declines. For instance, the U.S. Fish and Wildlife bird count, she notes, "has been going on only since 1966. That's just twenty-eight years, which is not a long time. The results so far show a decreasing trend, but it may be that we're simply seeing a dip within a larger natural cycle of up-and-down movement. We've known for years that animal populations go through natural ebbs and flows. We've known that ruffed grouse and showshoe hares were cyclical, for instance, because we've hunted them for food. But nobody eats warblers, so no one has paid a lot of attention to whether they were cyclical until recently. It would be more surprising to me if songbird populations weren't cyclical than if they were. So it may be that this apparent decline is partly an artifact of where things were in the cycle when we started collecting data."

The land Carol is using for her study is outside the Umbagog National Wildlife Refuge, in a strip of forest called the Thirteen Mile Woods, about five miles south of the lake and five miles north of Don Roberge's camp. It's bordered by a beautiful stretch of river that offers some of the state's best trout fishing and whitewater canoeing. On the side of the river opposite Carol's research area, the Thirteen Mile Woods is protected by a scenic easement that prohibits logging within 500 feet of the water's edge. Once owned by James River Corporation, and by the Brown Company before that, the Thirteen Mile Woods was bought by John Hancock Timber Resources Group (a division of the financial services firm) in 1993 as part of a 238,000-acre acquisition in northern New England. Hancock has said it bought its land in the Northern Forest not for speculative development, as many at first feared, but for long-term forest management; conservationists hope that the scenic easement will ultimately be extended to this side of the river as well. Carol's other study site, the regenerating clearcuts on Nash Stream, was owned by Diamond International and had been cut hard before the 1988 sale. It is now part of a state-owned reserve where limited logging will take place. In both places – the Thirteen Mile Woods because of its proximity to the refuge, and Nash Stream because the new state forest is planned to be a model of mixed-used management – the question of how cutting practices affect wildlife promises to be an issue.

Carol hopes that her breeding assessment technique will help provide an answer to this question. A more efficient, accurate way to track songbird breeding success would help researchers find specific causes for declines such as those recorded over the last twenty years; it would also enable scientists to evaluate precisely how different logging practices affect different songbird species, and, ultimately, provide landowners specific guidelines on how they can cut trees without crippling habitat.

Carol hesitates, however, to raise expectations too high. She knows how difficult it is to transform the findings of conservation biology into good forestry practices – or, to look at it another way, how easy it is to make poor decisions based on apparently unambiguous findings. As an example, she cites the long-held assumption among ornithologists that higher densities of birds meant better quality habitat. The problem with that assumption, she says, is that biologists have traditionally gauged bird densities primarily by counting singing males – even though, as she puts it, "counting singing males tells you nothing about whether they're trying to breed, much less whether they're doing it successfully." Recent findings, for instance, suggest that in the yearly competition for nesting sites, the best breeding habitat is occupied by breeding pairs, which take up more space than bachelor males do, because a breeding pair's territory must support not just a male, but a female and fledglings as well. The successful breeding pairs thus push the bachelor males into suboptimal habitat, where they live in denser numbers than the breeding pairs – singing all the while, and appearing, as it were, quite dense. In other words, taking singing males as an index of overall population health and density is something like estimating human reproductive vitality from a census of fraternity houses instead of family neighborhoods.

"This has some major implications," says Carol. "From a habitat protection standpoint, if you're trying to decide between protecting two different pieces of land and you protect the one with more singing males, you might be protecting the inferior habitat." It might have similar implications regarding an effort to evaluate the impact of two different logging practices; if you counted singing males, you might actually conclude the more compromised breeding habitat was the better for wildlife. This is one reason she's trying to develop a method that evaluates breeding success, not just numbers of singing males.

These ambiguities humble any attempt to claim conclusive knowledge. As Carol points out, "The fact is, we still know very little about how animals – from salamanders to moose – actually use the landscape, or how that relates to what has happened on the landscape. The whole thing is very, very complex. And it's *always changing*, whether people are there or not. People are *crazy* if they think there's some ideal balance of nature out there we can strive for or return to. The forest is not this static primeval wilderness that some would like to make it out to be. There is *always* a dynamic."

She shakes her head and starts to reach for another sandwich (this is her third; she talks a little more enthusiastically with each one), but leaves it on her plate as she leaps to the next thought.

"That's why the questions I find fascinating are the ones about the working landscape. What critters is that supporting? Can we make that working landscape more acceptable to more species without sacrificing anything horrendous on the production end? That's what interests me. We absolutely must have permanent biological reserves. But reserves aren't going to do it all, not in New England. We've been here too long. There are too many of us. We've got to find a workable way to co-exist."

5

Lakeside

Leo Roberge backs his boat trailer down the old Brown Company steamer landing on the Androscoggin, about a mile downstream from the lake. It is the first week of November, between the two halves of the split duck season, so we load no guns into the boat, only life jackets and binoculars. Leo handles the trailer as deftly as you'll ever see one handled: one quick curving reverse and the 16-foot bass boat is afloat.

The calm water here above the dam appears black under the pewter sky. The near side of our small wake reflects the color of the sky, then turns clear as the wavelets wash over the ice lining the shore. Winds have stripped the hardwoods bare, leaving the upper slopes of the hills gray. Snow covers the mountains to the east and south. We are glad it is calm, for with the temperature in the low twenties, even the breeze created by our trolling speed cuts through our coats.

Cold or not, we feel lucky to be here, not only because it is beautiful and immensely peaceful (for we are the only ones out), but also because Leo had so clearly been tempted, when we met him at lunch, to spend the afternoon hunting deer instead of showing us around. But with a smile and a shrug of his shoulders he had said goodbye to his companions and swept the snow off the boat and hooked it to his truck, and driven down here and put in. Once on the river, standing at the center console of the boat, he smiles. His brown eyes tear from the cold. The November wind has knocked the wild rice along the shore down to little more than strawlike grasses. It lies flat, held in the tension of the water's surface.

"I guess it was about the mid-seventies, maybe a little earlier, that Armand started us on the rice and boxes," Leo says. "He used to do what he could on his own, but then we started using the hunting club to raise money. Game banquets and so on. We'd take the money and buy bushels of rice. You just throw it out and let it sink into the mud and it plants itself. Now it's everywhere. It's spread all up and down the river.

"You see that little clearing there?" he says, pointing at the shore. "That's where I first started hunting. We had a blind in there. I'd sit with my grandfather, Alfred. My mom's dad. As a boy it wasn't too exciting. You know, 'Don't move, don't move, we gotta wait quiet for the ducks.' But as I got older I enjoyed myself a lot more. He taught us a lot.

"It was the same way with my son Keith. He's been out here hunting with me since he was three years old. I'd put him on my back with his little pop gun. First few times he blew it, couldn't wait for the birds to get close. I had to teach him, 'No no no, you wait till *Daddy* shoots before shooting that pop gun.'" He smiles, remembering his son as a boy.

As we move upstream, the river channel wanders among islands and fingers of land separating us from the larger expanse of the lake, in and out of backwaters and ponds and inlets. Here where the Androscoggin begins to define itself, the channel it cuts in the muddy bottom shifts with changes in flow and season, and with three of us in the boat, Leo is worried about running aground. Repeatedly we feel the boat pull, then surge ahead as the propellor grinds through mud, churning up a dark soup.

Duck blinds made of brushy branches are spaced every forty yards or so along the shore. "Up here at the head of the river is where the best hunting is," says Leo. "There's only a few places on the lake that are still good – Pot Luck Cove, Milan Meadows, here, a couple places on Leonard Pond. Which of course has the eagles now, so you can't hunt there. But everybody here knows just how the birds come in. They're taken by the same people every year.

We used to have a lot of hunters come up here. Now they go other places looking for more birds. Here on the lake, you have your diehards. No matter what happens, whether the season is good or not, we're out here. Some people think, 'I can't shoot my limit, I won't go.' But there's some pleasure in just going out here and doing it.

"But the birds are down. Blacks used to come down here in hordes of hundreds. I mean, you'd sit here and look at the horizon and see a big black V, just black ducks coming in. You couldn't go wrong."

Now, Leo says, you're lucky to see more than ten at a time. The census numbers on black ducks show that this is not, unfortunately, the distorted memory of a man recalling his youth. As recently as the late 1960s, black ducks far outnumbered any other duck species in the eastern half of North America; one authoritative work written in 1967 notes that "In spite of being heavily shot at, the Black remains plentiful and the sad tale that can be told regarding the diminution of the erstwhile numbers of many other species does not apply in the case of the Black." Since then, however, the ornithological literature records a precipitous decline. Studies with titles such as "The Black Duck Needs Help" (1982), "The North American Black Duck: A Case Study in 38 Years of Failure in Wildlife Management" (1983), and "Black Duck Decline" (1987) track the nosedive. Several times during the 1980s the bird was listed as a federal Species of Special Concern, one step short of threatened status.

Leo and his friends didn't have to read these studies to know that duck populations were dropping; the black clouds that had once swept onto the lake each November had become wisps.

"It was about fifteen or twenty years ago we started noticing," says Leo. "We said, Geez, we gotta start doing something here. That's when we started in heavy on the rice and boxes. Now the rice has started to seed itself, and some of it washes down, and as the years go on it keeps going down and downriver, and the ducks just love it, all of them. Your wood ducks, your black ducks, your mallards. Your teals. Even your songbirds. Sometimes you'll find a deer eating the stuff along the shore. It's good for all of them."

"There's a duck," says Leo, pointing over the bow. "A black. Low down there. You see it?" And finally we do. Three or four hundred yards out, a small black shape is moving fast across the lake, just over the surface. It shoots behind a point.

We're on the open lake now, and before us the water stretches to Maine. To our left is a long island separating Leonard Pond from the lake's open water to the east. Leo follows the island's shore northward. Over the island we can see the tops of the tall trees in which the eagles nest. But Leo is looking east, to the open water, where a set of broken black lines hold still on the lake's undulating surface.

"Something out there," he says. "Some blacks or some pintails." He pushes the throttle forward and shouts over the motor and wind.

"One year we came up here during the season and almost the whole lake was frozen. Except there was a break along the northern shore, and it was open at the other side over there. See that ridge on the other side?" He points at Maine. "The whole thing was covered with oldsquaws. They're white and brown birds, got a long tail in the back like a pintail. The migration was down and you could look on that ridge, and all you saw was a white cloud. It was all birds. So we followed the break over there and we all got our limit, three birds. Then we just sat there and watched them. They kept flying back and forth. It was *beautiful.*"

We're at the north end of the island on our left now. Leo skirts the head of it and continues north up into the channel formed by yet another island, leaving the eagles unseen somewhere behind us. On the island on our right are plastic duck boxes the color of band-aids. Ducks Unlimited used to sell them black to make them less conspicuous, Leo tells us, until they discovered that the black boxes got too hot. The brighter color is easy to see among the grays and whites and greens of the shorelines.

As the island on our right ends, the shore on our left bends in front of us and heads out into the lake. Near the end of this point, about a hundred and fifty yards distant, is a broken line of black: *birds.* We all see them this time – three dozen or more, gathered in two rafts just off the point.

"See there!" says Leo. "Them birds. Oh there, oh yeah. See that, those are all birds." Two are splashing, washing themselves. "Oh them birds. And we're out of season." He shakes his head. The two splashing ducks take flight. "See those, what, a couple loons? No, mergansers. Fish ducks." As we approach, Leo, squinting, tries to make out what the remaining ducks are. When we are about a hundred yards off he says, "Ringnecks." Then a moment later, in a low voice, "No. Decoys. Somebody's decoys."

We are all slightly embarrassed over our excitement of a minute before. As we pull closer it's clear that Leo identified the decoys correctly, anyway, even from a hundred yards: They are well-executed ringed-necks, with a couple of pintails mixed in, arranged artfully just off the end of the point. As we round the point we surprise two more live pintails, which fly off. On the other side of the point are more decoys and a blind, empty.

Leo turns the boat around and heads south again. He slows the boat as we enter Leonard Pond. Ahead is a small strip of an island with two tall, dead trees. In the farther tree perch two eagles, one on a huge nest near the top of the tree, one on a branch about ten feet below. Their white heads are easily the brightest thing in the landscape on this gray day.

"Whiteheads," says Leo. "Two males. The females have brown heads, I think." For the first time all day, Leo, probably basing his sex identification on what he knows about ducks, has his natural history wrong. In eagles it is age, not sex, that produces the distinctive white head feathers. Gathered so close to the nest, this is almost surely the mating pair.

Leo approaches, idling slowly, as the eagles watch. "They're not too worried, huh?"

We pull closer, near enough to see the ruffling of the birds' feathers in the breeze. When we get within forty yards, the eagles lift and fly to the other tree. We have violated that law of wildlife observation that says if you change the animal's behavior, you have gotten too close. With winter closing in and the lake freezing over, we have cost them precious calories.

Leo slowly rounds the point and heads north again, toward the other tree. As we move upwater, Leo keeps the boat further offshore. He pulls even with the tree and cuts the motor. The boat drifts to a stop, starts moving back downwater. Leo stands staring at the eagles, who stare right back.

It's February, and Don Roberge is putting up lights in the garage at his camp. At 25 degrees, sunny and still, the day seems warm, for the temperature has been in the single digits or below for almost three weeks. Don works in a wool shirt with the sleeves rolled up. When he gets a hole drilled he feeds the wire through and pulls as much slack out of it as he can. Then he staples it in place and comes down the ladder.

"I want to show you something," he says. He leads us past the garage and up the drive. At the back is a big berm where he has pushed the snow from the driveway. In one place he has plowed through the berm and scraped the snow off an area of ground between the driveway and the thick forest beyond. In the cleared oval are several mounds of deer feed. Any time Don goes into the house for more than five minutes, he says, at least a half-dozen deer come to feed, at first shyly peeking through the trees and then coming into the open. Don doesn't ordinarily feed deer, and he's not feeding them to lure them in, for the season is over. He's feeding them because this winter has been one of heavy snow and cold temperatures, and with over three feet of snow blanketing the county, food is scarce. He says the deer eat 100 to 150 pounds of feed a week. He can only guess how many deer feed here: at least a dozen, he says, maybe more. The entire area, every square inch, is covered with thousands upon thousands of deer tracks.

"And look at this," he says. He walks into the trees. A trail packed by thin skis slips into the black spruce and poplar woods that lie between the camp and the river. The tracks are completely trampled with deer prints. "I cut three miles of cross-country trails in here," he says. "I built a homemade ski tracker, a little frame with skis on it I could tow behind the snow machine, and I put in ski tracks. It worked all right but I'm going to try to get a used one from the cross-country center. The kids love coming back here to ski. Hey, I love it. You can ski right down the river and all along. It's nice easy skiing.

"'Course Leo, he'd rather take off across the hills," says Don. He shakes his head. "*Got to* go off the trail. Goes over the road and off the trail and up the stream and over the mountain and back. That's *hiking*, not skiing. He took my cousin and his wife out skiing one day a couple weeks ago, all the way over the mountain. They're gone, five hours. They come back. I say, 'Are you glad you went off with my brother?' Me, when I ski I'd rather stick to trails. I'd rather shuffle."

It's been almost a year since we've been to the camp, and Don has done a lot in that time. He has cut the trails into the woods, built a four-season porch on the back of the house so the kids would have a place to play bumper pool and be away from the adults, and cleaned up the grounds a bit. He talks about trying to buy the thirty acres of wood between the house and the river, where he has cut his trails. They're too low to be developed, he says, so he figures he might be able to take them off the present owner's hands for a few thousand dollars. Don wouldn't do anything with the land, but he'd feel better knowing no one else could do anything with it either, and that he and Leo would always have a place to hang duck boxes and ski. Though the camp started out

54

as a place for the brothers to hunt from, it has more and more become a place where they spend time with their families. Don spent the majority of deer season here, "most of it," he says, "trying to get this one big buck." He hunted for at least parts of ninety days, starting in bow season, through muzzle-loader, rifle, and then bow again. He never got the buck. He says there was a good deer take in the state this year.

"There was also a lot of moose," he says. "More moose every year now. My mission for next summer, I'm gonna ride one. I am!" Crunching through the snow back to the camp, he tells us in a half serious tone his plan to drop from a tree onto a moose's back. He says he wants to paint a blue spot on the animal to prove to his friends that he'd ridden it. Don has never actually ridden a moose before, but after knowing him for a while we figure he is as likely as anyone to do so. "I should get quite a good ride, too," he says.

By the time we get back, Leo has pulled into the driveway. Don berates him, in friendly brotherly fashion, for being late. We have a date to go down to the river and look at some duck nesting boxes along the shore.

"I know, I know," Leo apologizes. He had to finish a project of his own, hanging cabinets in his kitchen in Berlin. He says he's feeling spread a little thin lately, what with work and the kitchen project and teaching hunter safety in the fall and trying all year long to raise funds for two hunting clubs.

"Oh, the clubs!" he says. "You ever tried to raise money for two organizations at once? Hello," he says, holding out his right hand like a panhandler. "Do you know about the activities of the Umbagog Waterfowlers Association? We work to conserve and develop breeding and migratory habitat for waterfowl in the Umbagog region? Would you care to make a donation? Thank you." He pulls back his right hand and holds out his left. "Hello. Do you know about the activities of the Androscoggin Fish and Game Club?"

"So there's that," he says, "and building the duck boxes, we've got about thirty we still need to put up from last year, and checking them, and work, and the hunting. Oh man. It's something all the time." He shakes his head, smiling.

"Well let's go," says Don. While the rest of us change into ski boots, Don gets a ladder from the garage and ties it to the back of his snowmobile. Leo, on skis, leads the way out the back of the driveway and into the trail network. Clothed from head to toe in camouflage hunting gear, he skis with a utilitarian gait that keeps him moving steadily along the trail. We follow him through the trees and then slide down a short slope onto a marsh about a hundred yards wide. Don follows on the snow machine.

A narrow island separates the marsh from the main stem of the Androscoggin. Both marsh and river are frozen solid and covered with snow. We ski first to the island to check a duck box nailed twelve feet up on a dead tree. The box is a tan plastic cylinder eight inches across and two feet tall with rounded ends and a four-inch hole toward the top. It is one of about five hundred boxes owned by the Umbagog Waterfowlers Association. Leo says about 80 percent of the boxes are occupied in a given year, and most of those pairs breed. Some eggs never hatch. So as not to disturb the ducks during the summer breeding season, Leo and Don often wait till the following winter, as they have done with these boxes, to check for signs that the boxes were used. As Don climbs the ladder, Leo tells us that they recently checked three boxes upriver and found a dead mother duck and six unhatched eggs in one, seven unhatched eggs in another, and in the third, shells showing that ducklings had hatched.

We all look up to see what Don will find. When he puts his hand on the box, a brown blur shoots out the front and up the tree. Don, startled, shouts and jerks away, almost losing his balance on the ladder. "Ya goddam squirrel!" he says. The squirrel chatters at him from the top of the tree, tail flicking. Don taps the box. Another squirrel bursts out and scampers up the tree. "Oh Jesus," says Don, laughing now. He knocks on the box to make sure no one else is home, then unhooks it from the tree and brings it down.

Don swings open the front of the box and lifts a round plastic tray from the bottom. The tray holds a dense bed of dead grass and feathers.

"This one was occupied," says Don. "Those squirrels were just using this stuff."

He digs gently through the nest and finds several wood duck feathers. They are exquisite, about an inch-and-a-half long with fawn-colored ends and delicate black-and-white striping through the middle third.

"That is pretty," says Don. "Nice to tie flies with, too." He digs further and finds more feathers, and then what he's really looking for: some bits of pale shell. This nest produced offspring.

The second box we check contains signs less certain. A shallow layer of litter lies atop the cedar. "Might have been something here, might not," says Don. He digs through the cedar and finds only two small feathers and no shell fragments. He and Leo decide that a bird had probably investigated the box and perhaps even begun a nest, but not fledged successfully.

The last box, back across the marsh on the main shore, shows no sign of habitation. It holds only the cedar chips the Roberges had put in there in the spring. Don cleans out the dead leaves that have blown in and replaces the tray. Then he closes the front half of the box, locks it, and screws the box back into the tree.

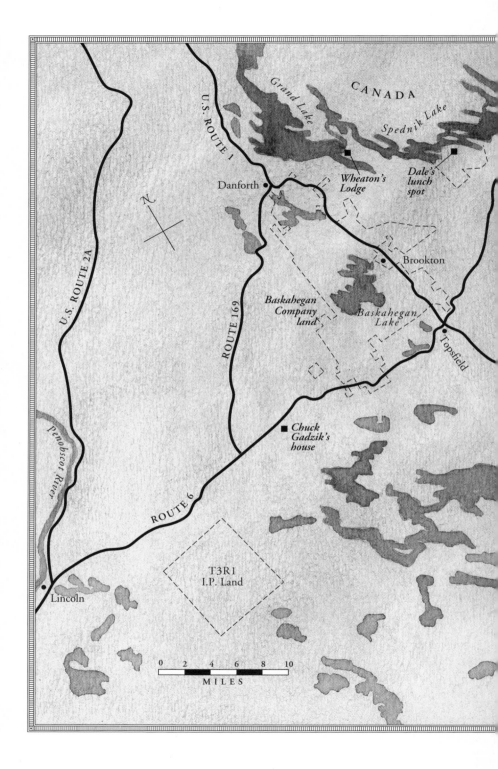

CANADA

Grand Lake

Spednik Lake

U.S. ROUTE 1

Wheaton's Lodge

Dale's lunch spot

Danforth

Brookton

U.S. ROUTE 2A

ROUTE 169

Baskahegan Company land

Baskahegan Lake

Topsfield

Penobscot River

Chuck Gadzik's house

ROUTE 6

T3R1 I.P. Land

Lincoln

0 2 4 6 8 10
MILES

Part Two

MAINE

—

*"If you don't know the ground,
you are probably wrong about nearly anything else."*
NORMAN MACLEAN

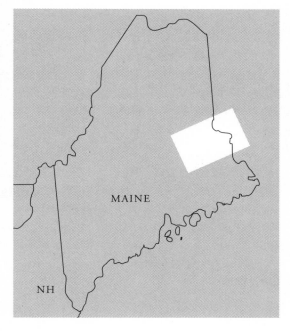

AREA OF DETAIL

6

When a Tree Falls

The pine didn't shudder until the blade of the ax had sunk more than a thousand times.

When the man had started chopping the tree an hour earlier, only three seconds would pass between each swing, before another chip would fly off the hardened steel and another seam of pitch would bubble out. Slowly he had deepened the cut into a wedge four feet across the trunk.

Now the swings were coming further apart as the man labored in the thin winter air. Under his canvas trousers and wool shirt, clammy sweat glued his skin to his long underwear. Fifty feet away, an ox stood staring into the snow. Occasionally the huge animal trembled, jingling the iron and leather rigging draped over its shoulders.

When the pine shuddered, the man stopped and wiped his face on his frozen sleeve. Stepping back through the snow, he looked up the trunk. Deep furrows of bark ran straight up unbroken to the first branches sixty feet above the ground; the crown, swaying slightly, was another hundred feet higher. The man walked around the trunk to check the pile of trees he had felled to cushion the big pine's fall. Dropping the smaller birch and spruce and fir trees into a lattice had taken two days of tedious work, and he wanted to make sure the pine landed squarely in the middle. Once the tree was on the bed he would cut the trunk into 16-foot logs, wrap a chain around each log, and pull them off with the ox. From there it was a short drag to the bank of Baskahegan Stream, where the logs would sit until ice out in April. He would leave the cushion trees to rot, even though several were more than two feet across. They were worthless and the winter was getting old. Before the river started to run, he had more big pines to drop.

He turned back to the pine and hefted his ax.

The tree had sprouted a hundred-fifty years earlier when a small, lightning-strike fire had rushed along the Baskahegan. The blaze took out the thick canopy of hemlock, spruce, and beech trees that had shaded the river bank for centuries. With the sun flooding the forest floor, birch, white pine, pin cherry, and poplar seedlings pushed up through the blackened ground. Struggling against the hardwoods and underbrush, the pines fought to dominate their columns of light. They grew a foot straight up every year, wasting little energy on horizontal branches. Within fifty years, a few of the pines had outraced the hardwoods and were spreading their branches in the sun. The shade in the forest deepened. The pines dutifully dropped cones, but their progeny couldn't compete against more shade-tolerant trees whose seeds were still present, deep in the soil. As the ground cooled beneath the pines, tiny fir seedlings spread through the forest. Underneath the firs, and eventually overtaking them, were the slower-growing and stronger spruce, interspersed with beech and maple. Eighty years after the fire, the fir was dying back and the spruce trees were 60 feet tall, fighting the hardwoods for space. As the decades passed, the spruce trees fattened under the canopy of the pines.

The woodcutter had found the stand of pines the previous fall, in September 1818, during a scouting trip from his hometown of Bangor. He had made his way on foot up the shore of the Penobscot River, turning east along Mattawamkeag Stream toward the uninhabited wilds above Baskahegan Lake. During the first part of his journey, he had come across clusters of freshly cut stumps where woodcutters had worked the winter before. By the time he reached the point where the north-flowing Baskahegan Stream runs into the Mattawamkeag, he was in virgin forest. Climbing a ridge and shinnying up a tree to get his bearings, the woodcutter immediately spotted the tall pines, their soft green crowns standing out in sharp relief against the reds and yellows of turning hardwood trees and the blue-green spires of spruce and fir. The pines were hugging the western bank of Baskahegan Stream five miles to the south; beyond that he could see the dark waters of low-lying Baskahegan lake.

The woodcutter returned to Bangor. He sketched a rough map and went to see the government land agent to pay for the rights to cut pine on twenty acres of uncharted land. He gathered his team and stocked up on supplies. In mid-December, after a foot of snow had fallen and the ground was frozen, he left town with three other men and two oxen. The men walked, and the oxen dragged a long sled on wooden runners. Under the sled's canvas tarp the men had stowed sacks of beans and flour, salt pork, a small barrel of molasses, axes, whetstones, wool blankets, and a few personal belongings. When they reached the Baskahegan, they built a twelve-foot-square cabin of balsam fir logs and filled the chinks with moss. Off to one side they erected hovels for the oxen. They made beds out of logs and lined them with fir boughs and blankets. In the middle of the cabin was an open fireplace for cooking and heat. A hole in the roof above vented the smoke.

Every day from dawn to dusk the woodcutters cut pine. They hunkered down in their cabin only during the fiercest storms. By the end of March, they had stockpiled sixty logs, 16 feet long and ranging from 30 to 60 inches across. When the first spring freshet of April started to move the ice floes downstream, the men broke camp and packed up the sled. At the river's edge, they carved an identifying symbol into the ends and side of each log. Then, using iron-tipped wooden pry bars known today as peaveys, they rolled the logs into the rising river. For the next six days they walked along the shore of the Baskahegan and the Mattawamkeag, poking the logs through tight spots with long pick poles. At the Mattawamkeag's confluence with the Penobscot, southeast of Mount Katahdin, the men turned their logs over to a river-driving crew working from double-tapered boats called bateaus. Two weeks later the drive landed at a water-powered sawmill on the bank of the Penobscot in Bangor. Millyard workers hauled the logs out with horses and separated them by the loggers' identifying marks. By then the woodcutters were back in town, planning how they were going to use the fee from the mill to finance the next winter's cut. They'd have to look even harder, they knew, to find another good vein of pine.

After the American Revolution, Boston's political leaders were eager to settle the vast forests of central and northern Maine, which belonged to the Commonwealth of Massachusetts. Settlements would produce tax revenues to pay war debts, lumber to build the young nation, and timber for masts and shipbuilding. Government surveyors started laying out square townships of 36 square miles, reserving three public lots in each town for the school, the church, and the minister's house. The new Americans' fear of wilderness was still strong, however, and the land agents had a hard time convincing settlers to move north into the wilds. The Commonwealth granted land to Revolutionary War veterans and to colleges, sold chances to potential settlers through a lottery (a three-dollar ticket bought a guaranteed one hundred acres and a shot at winning as much as an entire township), and transferred huge tracts to private entrepreneurs such as William Bingham, who bought two million acres along the Penobscot and Kennebec Rivers in 1793. Offering incentives for homesteaders who cut timber and built sawmills, Bingham and other businessmen managed to get a number of towns started. Most of the successful settlements were located along the rivers, however, and the sprawling interior was left uninhabited. Independent loggers ventured into the woods to cut trees, but they usually returned to town in the spring. When the Bangor woodcutters had their lucrative winter cutting pine in 1819, for example, they were working on land owned by Massachusetts.

By the time Maine achieved statehood in 1820, less than half the new state's land area was in private hands. In partnership with Massachusetts, the Maine legislature tried to bring some order to the process of settling its central and northern territories. Starting at the state's eastern border with Canada, the government surveyed the land, divided it into ten-township areas called "ranges," and tried to sell the first five ranges, in 200-acre parcels, to homesteaders. Settlement was still spotty, however, and in the 1840s the state started auctioning 500-acre lots to logging companies and wealthy families from Portland, Boston, New York, and Philadelphia. In this way, a handful of entrepreneurs – including the Pingrees, Lords, Coburns, Prentisses, Carlisles, and Philanders – accumulated millions of acres in Maine's north woods in the 1840s and 1850s. The "timber barons" collectively employed thousands of itinerant woodcutters and river drivers. Living in primitive woods camps for up to six months of the year, the loggers cut all the big pine and spruce trees and drove the logs down the Androscoggin, Kennebec, Dead, Penobscot, and St. Croix Rivers to the mill towns and shipyards. To facilitate the log drives, David Pingree and other large landowners obtained flowage rights and built dams on most of the rivers. With the profits from sawing lumber, the companies bought more land and built more mills.

By the middle of the nineteenth century, Maine was the biggest producer of lumber in the world. Pushed to maximize production, woodcutters refined their craft through the 1850s and 1860s. They replaced oxen with horses, axes with crosscut saws. They worked in teams, piling logs in concentration yards instead of leaving the wood scattered along the streams. As more men went to work in the woods, they built more livable camps. The days of the small woodcutter who found, cut, drove, and sold his own logs were over; lumbering had become big business. Settlement, however, reached its zenith in 1854. The sheer immensity of the the north Maine woods – a region larger than the combined areas of New Hampshire, Vermont, and Massachusetts – confounded the government's efforts, and the state took until 1880 to sell the last of its land. To this day, most of Maine's northern half remains virtually uninhabited. Hundreds of unorganized townships – tracts of land where there is no local government – are identified only by number.

Bangor's mills sawed 250 million board feet of lumber in 1872, the city's all-time peak production year. By then, Maine's primacy as a lumber capital was already fading. All the big pine and spruce were gone, there was increased competition from virgin forests and new mills in upstate New York and Michigan, and the Civil War had taken a tragic number of young men who had worked in the woods. (Among the Union states, Vermont and Maine ranked first and second, respectively, in soldiers killed per capita.) Moreover, many veterans who survived the war never returned to work in New England's rural areas, having been granted land in the more arable farmlands of the West and South, or having taken manufacturing jobs in the growing industrial cities. Over the next two decades, sawmills closed and dozens of lumber companies failed. Shorn of its pine and large spruce, much of Maine's forest by the 1880s consisted of balsam fir, hemlock, birch, maple, and small spruce trees. These trees were considered to be of little value for lumber, because sawyers and builders had gotten used to large, knotless spruce and pine logs that were easy to work.

Already, however, several new industries were emerging that could make use of the "worthless" species and the smaller trees that had been left by the timber barons. These emerging industries included leather tanneries, which extracted tannin from hemlock bark to cure hides; portable steam-driven sawmills, which could manufacture lumber deep in the woods away from the rivers; boxboard factories, which cut smaller and crooked pine and spruce trees for making packing cases; veneer mills, which stripped thin layers of wood from yellow birch and other hardwood logs for finishing fine furniture; and turning mills, which used lathes to manufacture bobbins, spools, dowels, furniture parts, and hundreds of other products for an increasingly industrialized and prosperous society.

The technology that would have the most profound impact on Maine's forests had been invented in Pennsylvania in 1867, when a chemist figured out that a compound called caustic soda would dissolve wood into pulp, which could then be cheaply made into paper. One of the most important advances of the industrial age, reducing as it did the need for expensive rags to make paper, the pulp revolution led to a pattern of land use and ownership that would dramatically alter Maine's economic and physical landscape. From 1867 to 1872, as every newspaper in the country switched over to newsprint made from wood pulp, a number of small paper mills were built in Maine. At first these mills used poplar, a fast-growing species that had thrived after Maine's big timber was gone and the farms were abandoned. With the discovery in 1888 that sulphite, a derivative of sulphur, could make pulp from softwood trees, fledgling companies such as Eastern Manufacturing Corporation of Bangor and International Paper of New York started buying huge tracts of land in Maine from the failing lumber companies and farmers, and building paper mills along the rivers. Tree size no longer mattered. A tree was worth money as long as it held enough fiber to grind into chips, dissolve into pulp, and press into paper. Eastern Manufacturing Corporation built Maine's first commercial wood pulp mill in Bangor in 1890. In 1899, Great Northern Paper Company started building a huge complex at the convergence of the east and west branches of the Penobscot River. Around the mill the company built Millinocket, which came to be known as the "City in the Wilderness." By 1910, pulp and paper mills were operating in Rumford, Woodland, Bucksport, Westport, and Jay.

The last big pines in the watershed of Baskahegan Lake were cut in 1845, but loggers were already turning to the area's abundant spruce. That year, two lumbermen from Bangor founded the Baskahegan Dam Company at the outlet of the lake to drive logs down the Baskahegan and Mattawamkeag Streams to Bangor.

Other lumber companies working in the area were already sending wood down the St. Croix River to Calais, on the Canadian border, or down the Machias River to Machias, on the Atlantic coast. For the next sixty years, the area's wood industry roughly paralleled the boom-and-bust cycle of the rest of the state: long lumber, boxboard, hardwood, veneer, and tannery mills came and went. Woodcutters and farmers slowly settled the towns of Danforth, Brookton, Topsfield, and Forest City.

In 1911, the area's dominant lumberman, Henry Putnam, bought the Baskahegan Dam Company and flowage rights to Baskahegan Stream, and, from other sources, 29,000 acres of land. Putnam already owned thousands of acres of land and several mills. His empire, however, was not to last. In 1920, he sold the Baskahegan Dam Company, along with much of his other land and mills, to five investors headed by Gerrish Milliken of New York.

Gerrish Milliken was president of a rapidly expanding textile empire that his father had started as a dry-goods business in Portland, Maine. Like many wealthy families who have made the Maine woods part of their portfolios, the Millikens had a hands-off approach – they were looking for an investment, not a lifestyle. Although he put up most of the $890,000 to buy the Baskahegan Company, Gerrish Milliken left day-to-day operations to others. Chief among these was a legendary woods boss and log broker named John Kelley, who was to own 20 percent of the Baskahegan Company. A larger-than-life figure with flaming red hair, "Dynamite" John Kelley, so-named because of his fondness for using explosives to break log jams, had run river drives and logging camps for Great Northern and the Eastern Corporation for fifteen years. By the time he met Gerrish Milliken in 1919, Kelley was one of the Penobscot region's foremost wood brokers and owned interests in hundreds of thousands of acres.

By then, wood was getting hard to find. The boxboard boom had peaked in 1909, and the last virgin spruce and pine logs had been floated to Bangor in 1905. Although the paper companies had been replanting some areas, the decades of cutting smaller and smaller trees for pulp were taking their toll. The hard cutting had been exacerbated by three natural disasters: a hemlock blight at the turn of the century, a rash of wildfires from 1904 to 1908, and, most devastating of all, a massive outbreak in 1912 of spruce budworm, a cyclical pest that kills fir and spruce trees by eating the tiny buds of new branches.

Before the new incarnation of the Baskahegan Company was a year old, John Kelley had negotiated a twenty-nine year pulpwood contract with the Eastern Corporation. The agreement called for Eastern to buy 500,000 cords of spruce, fir, and hemlock pulp from the Baskahegan Company's land. Eastern would pay Baskahegan five-and-a-half dollars a cord, in advance, every year. Aware that he was in a race against the budworm, Kelley stipulated in the contract with Eastern that an acre would not be considered completely cut until there were fewer than two cords of the target species left standing. (Previous owners had cut the land hard, and this clause was supposed to prevent Eastern from complaining when there were only spindly trees left to cut.) Kelley also convinced Eastern to pay the Baskahegan Company an overall advance of $250,000.

In the winter of 1920, loggers working for Eastern cut 22,000 cords from Baskahegan's land. They felled the trees, bucked them into 4-foot pulp logs, dragged them to the rivers with horse-drawn sleds and steam-powered tractors, and drove them downstream to Eastern's paper mill in Bangor. The loggers built roads and camps, which they used until all the marketable wood within reach was gone, and then they moved on. The Baskahegan Company earned a healthy profit the first year and bought more land. The next three winters produced even higher returns as the forest was turned to pulp.

The Baskahegan Company was not alone in the way the owners viewed their woodlands. All across central and northern Maine, entire townships were being reduced to thickets as the paper companies boomed and the woodcutters raced to stay one tree ahead of the spruce budworm. During the 1920s, the loggers and the budworm were removing twice as much wood as was growing; by some estimates, the pest alone killed half the state's spruce and fir stands.

In 1927, Eastern Corporation realized it had made a monumental error in the Baskahegan deal: At best, the Baskahegan Company's land might have had 250,000 cords when the contract was signed, only half what Eastern had agreed to buy. Eastern took the case to court and terminated the deal. In 1929, the Baskahegan Company started cutting the pine and cedar and hardwood trees that had not been included in the contract and sawing the wood at its own lumber mill. That fall, the stock market crashed.

While the paper business weathered the Great Depression comparatively well in the still heavily forested southeastern and midwestern states, New England's mills suffered as they scratched for wood in the ravaged countryside. Kelley tried to keep Baskahegan's revenues alive by hitting the few remaining stands of birch and pine, but his efforts made little difference, and he resigned in 1935. Three years later, the Millikens hired timber appraiser George Carlisle to assess what was left of the timber on their land.

"Thickets, thickets, thick, thick, thick stuff everywhere."

That's how George Carlisle would later describe how major pieces of the Baskahegan Company's land looked when he toured the ownership in 1940 with Gerrish Milliken and his son Roger. By then the Millikens owned 83,000 acres in northern Washington County, mostly in a polygon in the towns of Brookton, Danforth, Topsfield, Forest, and Kossuth. Viewing the jungles of raspberries, poplar, and scrawny gray birch from his motorcar, Gerrish Milliken was so discouraged he wanted to sell. The family's textile empire had made the Millikens wealthy, and there seemed no need to hang on to what had become a failed investment. But his son Roger convinced him instead to borrow from another Milliken family business to pay the taxes, and to hire a new manager to "tie down all the hatches" and let the land grow back. As absentee owners, however, the Millikens couldn't monitor the results of this decision. The local man they hired to keep an eye on the land spent the next ten years selling wood and pocketing the income.

Nevertheless, by 1951 parts of the Baskahegan land had grown back enough to justify another pulpwood harvest. The family rehired John Kelley, who immediately negotiated a contract with the St. Regis paper company of Bucksport, Maine. Armed with new motorized equipment, including tractors, trucks, and chainsaws, which had been developed during World War Two, St. Regis' loggers systematically cut every tree over 5 inches in diameter. (After Kelley and St. Regis had been through an area, one local woman quipped, "You couldn't find a stick of wood big enough to hit a dog with.") But the contract made money for the Baskahegan Company for a few years, until Kelley died in 1956.

In 1957, Roger Milliken, who had succeeded his father as president of the Baskahegan Company, sent a lawyer named Harold Kennedy to Maine to evaluate the firm's prospects. The situation looked grim. A handshake deal John Kelley had made before his death with a lumber company called Samson and Adams had gone sour, several towns were assessing unusually high property taxes on Baskahegan's land, and only a few pockets of mature timber remained. Kennedy promptly straightened out the tax problems, and he sued Samson and Adams. He also recommended that the company hire Fred Kinney, who had been John Kelley's assistant, as woods manager. Milliken agreed, but he still believed that the best way to increase the land's value was to let it sit and recover. He would not authorize any major logging contracts.

Little was done in the woods until 1960, when Kinney asked Milliken to allow loggers to thin a few mature stands of fir and birch that had escaped the earlier harvests. Kinney had been visiting tree farms and attending forestry forums, and had come to realize that cutting pulpwood on thirty- to forty-year cycles would never increase the long-term value of the land, because the trees would never have a chance to grow into valuable sawlogs. Along with many of his peers, he believed it made more sense to enter the woods more frequently and with a lighter hand, thinning the mature timber and culling out diseased or deformed trees. This sounded sensible enough to Roger Milliken, who agreed to give it a try. Under Kinney's close supervision, the first selective harvest made the company a small profit. In the next few years, Kinney contracted for several other small cuts.

In 1964, Curtis Hutchins of the Dead River Company of Princeton, Maine, invited Kinney and Kennedy to Dead River's tree farm in the town of Lakeville. The Baskahegan officials were impressed to see how Dead River's foresters were using careful selective cuts to accelerate growth of spruce and pine, and they arranged for Hutchins to meet with Roger Milliken. Hutchins convinced Milliken that a management contract with Dead River would meet the Baskahegan Company's objective of improving the long-term productivity and value of the land; Dead River, Hutchins said, had the same goal for its own holdings in the area. Ultimately Milliken agreed, especially when Dead River insisted that Baskahegan retain an independent consulting forester to oversee the contract. For this task, Kennedy hired Zebulon White and David Smith, two highly respected faculty members of the prestigious Yale School of Forestry.

Confident that the Baskahegan Company was in good hands, Roger Milliken signed a five-year management contract with Dead River in 1967. A half-century after his father had first bought the land, Milliken had finally found a way to invest in the property's value rather than simply cash in on its assets.

"We were not looking for a quick buck," he would tell his son some years later. "We were going to take the long view."

7

Imprints of the Past

On a warm Friday morning in May, Chuck Gadzik sits at his kitchen table drinking coffee. The early sun shines into the kitchen through a glass porch, casting pools of light on the hardwood floor. Near the center of the room stands a refurbished wood-burning cookstove. An L-shaped counter on the wall across from the porch holds a 12-inch black-and white television, an electric range, and a microwave oven; a dishwasher is tucked underneath. In the adjoining living room, the Canadian Broadcasting Company's 6:30 A.M. news murmurs from a radio. A magazine basket next to a white cotton couch overflows with dog-eared copies of *Consumer Reports, Utne Reader,* the *Journal of Forestry.* Several Winslow Homer reproductions are carefully positioned on the whitewashed walls.

Chuck bought this house nine years ago, in 1984, shortly after he began working as chief forester for the Baskahegan Company. He was taken with the old farmhouse's solid granite foundation and its seventy acres of fields and woods. Located in the tiny community of Carroll, the house was only fifteen miles from Baskahegan's office in Brookton. He started renovating right away, enclosing the porch, rebuilding the kitchen, erecting a barn, replacing windows. He married in 1986, and for several years his wife, Chris, helped out with the remodeling. Chris and Chuck separated in 1992, but her touches remain: a shorter pair of cross-country skis standing next to Chuck's, bright cushions on the couch. Since the separation, Chuck has been thinking of selling the house, but he knows he would lose a lot of money.

73

Chuck looks around the room. "I could keep working on this house forever, but I'll never get my investment back," he says. "There's no way I'll ever sell for more than I've put into it." Chuck wears a faded blue dress shirt and a gray baseball cap. Along with his soft, clean-shaven face, the cap makes him look younger than his thirty-six years. He gets up from the table and looks out through the glass porch. Below the house, a wide green field slopes down to a dark band of trees two hundred yards away. Chuck is describing how he wants to dig a fish pond to catch runoff from his well when the phone rings. He answers it on the kitchen extension.

It's not unusual for Chuck to get business calls at home, even at 6:30 A.M., and he has been expecting this call from a wood buyer at Champion International Corporation. He goes to his study to finish the conversation.

Baskahegan has been working out a new contract to sell spruce logs to Champion's sawmill in Costigan, 60 miles to the southeast, and the wood buyer wants to pin Chuck down on price and volume. The negotiating, as always, started from a tough point: Champion would like to lock in a deal to buy high volumes of wood at low prices; Chuck, on the other hand, prefers to sell low volumes at high prices. Chuck is confident they'll reach agreement, but the buyer is pushing as hard as he can.

Champion, of course, could just refuse Baskahegan's terms and look for wood elsewhere. That's what might have happened in the early 1980s, when there was a wood glut in Maine and the mills more fully dictated prices and terms. That glut is long past, however. Now, with demand outpacing supply, it's more of a seller's market. Champion can't afford to alienate landowners, especially those like Baskahegan who have good softwood to sell. Every year, the Costigan mill needs thousands of truckloads of spruce and balsam fir to make 75 million board feet of "studs," two-by-four and two-by-three-inch boards used to frame houses. Although Champion owns 730,000 acres of timberland in Maine – seven times Baskahegan's holdings – the corporation does not have enough mature trees to provide more than a fifth of the logs used at Costigan; the company buys the rest on the open market. The stud mill, moreover, accounts for only part of the company's overall need for wood. Champion's modern paper mill in Bucksport uses 900 tons of softwood and poplar every day to make paper for seven of America's largest magazines, including *Time, Sports Illustrated,* and *Cosmopolitan,* and for several mail-order giants; the paper needed for L.L. Bean's catalogues alone keeps one paper machine running for two months every year. Just over half of the 150,000 cords used every year at Bucksport comes from Champion's own land, so it is always shopping for wood. "If you own land and you want to sell wood to Champion," says one senior company official, "we'll make it possible."

Champion is one of twenty manufacturers that regularly buy wood harvested from Baskahegan's land. Approximately three-quarters of Baskahegan's annual cut is lower-grade hardwood and softwood trees. Chuck sells these species as chips to Georgia-Pacific in Woodland, which makes paper and "manufactured lumber" such as waferboard; and as logs to James River Corporation in Old Town and Great Northern-Bowater in Millinocket, both of which make paper. Most of the balance of the cut is in spruce, fir, white pine, and hemlock sawlogs, which are used for making lumber at Champion's mill in Costigan, Crete Manufacturing in Dover, and J.D. Irving in New Brunswick. Baskahegan also sells high-quality hardwood logs to specialty manufacturers, including a local shop that makes canoe paddles of maple and a factory in New Brunswick that makes tool handles of ash. Occasionally Baskahegan sells a load of premium birds' eye or curly maple logs for export to European furniture makers. In all, Chuck oversees harvests of some 35,000 cords of wood a year, which gross well over two million dollars.

Chuck returns to the kitchen and refills his mug. "There's always this tension between the resource and the mills," he says, sitting back down at the table. "Champion made the mistake through the 1970s and 1980s of hammering their own spruce volumes just to feed their stud mill. Strategically it was a very dumb thing to do, because eventually they hit the wall. Now they're at the mercy of people like me.

"We sit down at the beginning and negotiate terms," he says of the contracts with Champion and other buyers, "but the minute we stop talking, things start to change. I have to have a sense of how much wood we're producing and what kind, but that's not something I can calculate at some point in time and then walk away from. Because if we get locked in to producing just a certain amount or kind of wood to satisfy one mill's requirements, we lose the flexibility to make the kind of forestry decisions we want to make."

When selling spruce and pine logs, for example, Chuck commits only to low volumes. Then if one of the mills falls short, he can renegotiate. He signs contracts with specific terms, but these are generally used more as benchmarks than as ironclad guarantees. Growing and harvesting trees, he says, is simply too unpredictable a business. He credits the Millikens' past troubles in part to their inability to accept these uncertainties.

"In the Millikens' manufacturing world, you give the exact quality and the exact delivery your customer wants, and everybody stays happy," Chuck says. "Well I'm kind of locked into this death grip with our customers. If I give them everything they want, I may be a good supplier, but I'm also a stooge. So I have to play hardball with them. The key issue for us is what happens to wood values. Will they continue to grow with inflation? Will they do better?"

Commerical timberland owners sell wood in two ways: by stumpage and by contract. With stumpage sales, the landowner sells standing timber to a logger. The logger builds roads, cuts the trees, sorts the logs by species and grade, and sells the different kinds of wood to a variety of mills. His gross is the difference be-

76

tween the price he gets at the mill and the price he paid for the uncut trees; his net comes after he subtracts roadbuilding, equipment, fuel, insurance, and labor costs. To make a profit under these terms, the logger often has to move fast and cut hard. That's why Baskahegan prefers to sell most of its wood by contract instead. Chuck agrees to deliver a certain volume to a mill, then pays a logging contractor a per-cord fee to cut and truck the wood. These agreements provide year-round work for seventeen loggers and logging equipment operators, three roadbuilders, and seven truck drivers. In addition, an independent surveyor and a consulting forester work approximately half-time on Baskahegan's land. The Baskahegan Company itself has only four employees: chief forester Gadzik, assistant forester Paul Cushman, a part-time office manager, and Roger Milliken, Jr., who is president of the company.

The Baskahegan Company's principal objective is to improve the long-term value of its land. Chuck believes there are three principal ways to accomplish this goal: maintaining low overhead, investing in good roads, and cutting lower-value wood whenever possible. There is no mortgage on the land, so the only major annual operating expenses are salaries, taxes, and roadbuilding. Under Maine's Tree Growth law, which grants reduced property taxes to landowners who leave their land in forest management rather than developing it, the company pays about $100,000 a year in property taxes. Roads cost an average of $180,000 a year. In an era when many foresters rely on digitized maps, satellite images, and computerized inventories, Baskahegan is decidedly low-tech. The Baskahegan boat, an important tool considering the company owns more than 100 miles of shoreline, is a dented, 12-foot aluminum craft with a 1956 Evinrude motor. While paper company foresters cruise around in big four-wheel-drive pickups emblazoned with corporate logos and equipped with two-way radios, Chuck plies Baskahegan's gravel roads in a company-owned Toyota station wagon.

The more infrastructure you try to support, the more cash you have to generate, Chuck explains, and the only way to provide that cash is to take it out of the land. That's something the Millikens don't want to do anymore.

"The land appreciates depending on how much pressure you put on it. For us, by having relatively conservative harvest levels, particularly in the quality categories, we're increasing value in three ways: The trees are growing. We're shifting the growth to the higher quality product. And the wood products themselves are appreciating in value. That's a diverse portfolio. It's not producing six percent a year in terms of cash flow, but the value is increasing. It's like putting money in the bank. You wait for the time until it is a big fat account, and then you can start drawing from it."

Chuck Gadzik grew up comfortably in Storrs, Connecticut, the son of a public-works engineer and a stay-at-home mom. During high school he grew increasingly uneasy with suburban life, and he started thinking about studying forestry in Maine. Chuck had spent several vacations and one memorable summer in the north woods and knew he wanted to work outside. In 1975, he enrolled as a forestry major at the University of Maine at Orono.

Chuck studied forestry for the same reason that attracts most forestry students: he was passionate about the woods. Once forestry majors complete their education, however, the harsh reality of cutting trees for profit sets in. "Then," Chuck says, "they go one of two ways. They either are very intrigued by the industry that they probably had no idea existed, or they just reject it as some outrageous horrible thing."

Chuck was intrigued. In the summer of 1979, right after graduation, he traveled to Sweden to study that country's sophisticated logging techniques. When it was time to return in the fall, he decided instead to tough out the winter in Sweden as a logger. During the short winter days he cut trees and piled logs, stopping occasionally to thaw out his feet and hands in small warming huts, where he listened to the Swedish woodcutters boast and tell stories. Chuck returned to Maine in the summer of 1980. He worked for a while as a consultant, demonstrating the felling techniques he had learned in Sweden, and spent eight months on a logging crew in western Maine. In 1982 he got a job with the Cooperative Forestry Research Unit at Orono, where he studied how forests reacted to thinnings. During this time, he met as many people in the industry as he could, trying to catch up on the practical business education that he had missed in school.

Chuck was at Orono in the fall of 1983 when he heard that the Baskahegan Company was looking for a new forester. The contract with the Dead River Company had ended badly, and the Millikens wanted a woodlands manager with both academic and practical experience to help them regroup. They were finding it hard to recruit a veteran to remote Brookton, however, where the nearest supermarket was 30 miles away. This was precisely the chance Chuck had been hoping for. While he had virtually no experience working with logging contractors and wood buyers, his connections in Maine and experiences in Sweden got him an interview and, ultimately, the job. He started work in December 1983.

As Chuck has come to know Baskahegan's land, he thinks often of John Kelley and other past managers who have had responsibility for the forest. It's hard not to, for every day he confronts the results of the decisions they made. When he looks at the woods, he sees most clearly the influence of the Dead River Company.

"Our efforts now are hampered by the imprint that Dead River left," Chuck says. "They had a lot of people whose careers were formed in the late 1960s. They wanted to do the right things, but their approach just wasn't adding up anymore. Everything they learned was going out the window, and they couldn't recognize the change."

In 1967, after three years of negotiation, the Baskahegan Company signed a five-year management contract with the Dead River Company. The deal called for Dead River to cut 25,000 cords of wood a year from Baskahegan's land, half of the forest's annual growth. (By then, Baskahegan owned 100,000 acres of land. One-half cord per acre, per year, is a fairly standard rate of growth for average mixed softwood and hardwood stands in the Northern Forest. So about 50,000 cords of wood were growing on Baskahegan's land every year.) Dead River managers would inventory timber, negotiate with loggers, build roads, and follow the instructions given by David Smith, Baskahegan's forestry consultant, to cut mature balsam fir whenever possible; like many foresters, Smith feared another outbreak of spruce budworm, a misnamed pest that actually prefers fir to spruce. Dead River hired several three-man logging crews. Two men cut the trees with chainsaws, and the third hauled them out with a cable skidder, a machine that was just beginning to appear in Maine's forest, and which was essentially a heavy-duty woods tractor equipped with a massive winch for dragging wood. After the first two years, to the relief of Roger Milliken, Sr., Smith reported that Dead River was following through with the selective cutting methods they had promised. Baskahegan started turning a profit. In 1972, Milliken extended the deal.

Soon the trouble began. As foresters had predicted, an outbreak of spruce budworm started spreading across the state. Desperate to cut the fir before the pests got to it, David Smith told Dead River's managers to increase the annual harvest to 37,500 cords – a level that should have been easily sustainable, for it was only three-quarters of annual growth. Rather than spreading out the impact by thinning more acreage, however, Dead River started targeting Baskahegan's best stands and cutting out all the spruce and pine larger than 4 inches in diameter.

This return to harvesting by "diameter limit," which had been John Kelley's style, led the Millikens to question whether the Dead

River Company had changed its mind about the benefits of lighter cutting. But the Baskahegan Company itself was partly to blame. For some time, the Millikens had been sending Dead River mixed messages – insisting on keeping expenses down, while still demanding the highest quality forestry. The low point came when the owners refused to invest in new roads. To make a profit while paying the full costs of road construction, Dead River was forced to cut harder in each area they could reach.

The more direct causes of Dead River's change in strategies, however, were the ceaseless pressures of an industry in transition. When Dead River had started cutting Baskahegan's land in 1967, loggers were still using horses, skidders were rare, some pulpwood was still transported by river, and the industry was just pulling out of a long decline in capital investment. By the mid-1970s, however, the industry was rapidly modernizing. Fortune 500 companies had bought up all the state's paper mills and the land that went with them. New sawmills and manufactured-lumber plants were coming on line. Bulldozers were building truck roads into areas that, because of their inaccessibility to the rivers, previously had been cut lightly or not at all. Ever larger and more maneuverable skidders were pulling out higher volumes of wood. By mid-decade, some landowners were using whole-tree harvesting machines and on-site chippers to clear hundreds of acres at a time. As the budworm spread, cutting intensified. As supply rose, wood prices dropped. Harvesting costs remained high.

To stay ahead, Dead River started cutting as pulpwood some of the stands that David Smith had hoped to grow into sawlogs. In 1977, Smith went back to Dead River's Lakeville tree farm and found that the company had cut down all the showcase trees its foresters had worked so hard to cultivate. Smith convinced Roger Milliken that Dead River had changed its philosophy, and urged him to alleviate the pressure on Baskahegan's land by sharing the costs of roadbuilding and hiring a local representative to keep a closer eye on Dead River. These measures slowed the heavy cutting somewhat, and in 1979, Baskahegan earned enough to pay off the last of its debts. That same year, President Jimmy Carter settled a long-standing land dispute by providing funds to the Passamaquoddy Indians to buy tens of thousands of acres in northern Maine that the Native Americans claimed had been taken by illegal treaties. Much of that land had for decades belonged to the Dead River Company; after selling to the Passamaquoddy tribe, Dead River started to back out of the woods business. Baskahegan's last contract with Dead River ended in 1982. A year later, Roger Milliken, Jr., succeeded his father as president of the Baskahegan Company.

The younger Milliken wrote a book, *Forest for the Trees,* chronicling his family's company.

> Gerrish Milliken had bought the forest solely as an investment. For years, guided by John Kelley, and the prevailing pulpwood mentality, the Baskahegan cut the land hard for short-term profit – and paid the long-term price: accumulating debt with no resource to pay it off. With Dead River, the Baskahegan set out to have its philosophy implemented, but, once again, the owners didn't pay close attention. This is nothing new to Maine forestry. Charles Oak, Forest Commissioner in 1894 wrote: "Can it be a matter of wonder then, when the owner himself is so indifferent, that the operator – the person cutting under permit – is even more so, and that the lumber is cut away in a very wasteful and indiscriminate manner?"

Educated at Harvard and a student of eastern religions, Roger Milliken, Jr., was convinced that environmental and economic integrity were compatible. The themes that emerged most strongly in his 1983 book were the same on which he was determined to base his leadership of the Baskahegan Company: Stay involved. Keep your overhead down. Learn everything you can about how the business works and the forest grows. Above all, resist short-term profit in favor of long-term gain by accepting a simple truth:

"The Baskahegan Company," his book concluded, "has learned that its best interests lie with the best interests of the forest."

One of Roger's first acts as president was to try to find a forester who shared that conviction. He found him in Chuck Gadzik.

We eat breakfast at Daggett's Store in Topsfield, where dusty cans of Habitant Pea Soup are displayed alongside orange plastic ponchos and canvas work gloves. Chuck orders fried eggs and potatoes. He glances at the sports page of the *Bangor Daily News*, folds the paper back up, and strikes up a conversation with the only other customer sitting at the lunch counter, a bleary-eyed log truck driver who says he's been on the road for three days. The cook/waitress behind the counter joins in the discussion about wood prices and delivery schedules, calling both men by name and refilling coffee cups before they are half empty. After breakfast, while Chuck is gassing up at the Irving pumps across the road – the tiny town's only other retail business – a meaty man in a green-checked flannel shirt strolls over. After a few minutes of small talk, he gets around to asking if Baskahegan might sell or trade some crushed stone and screened gravel for his driveway. Chuck says he's sure they could spare a couple of pickup loads, and tells the man to call him later.

Chuck has lived in rural Maine for twelve years, but he's still a little wide-eyed at the easy commerce between neighbors that drives much of the region's economy. "We do two million dollars in business a year with mills and contractors, and most of it is verbal," he says, driving north on Route 1 toward Brookton. "We have to write contracts because of the lawyers, but it's the verbal agreements that matter. The people are so honest up here. You do what you say you're going to do. You just lay it out there."

The relationships can get dicey. Baskahegan owns about half of all the land in Topsfield, Kossuth, and Danforth, and virtually all of Brookton. The ownership is modest compared to that of the paper company giants, but Baskahegan's principal assets – roads, wood, and gravel – are indispensable to life in these towns, and proprietary lines blur easily. For example, Chuck allows loggers who contract with the company to cut a dozen cords of firewood for free. It's intended for personal use, but occasionally someone takes advantage of the company's largesse. This spring, assistant forester Paul Cushman told Chuck that one of the loggers was cutting oak for his father's carpentry hobby. Oak is uncommon this far north, and Paul wanted the trees left as seed trees. Chuck told Paul to keep his eyes open, but not to press the point unless the cutting got out of hand. He has learned that for every extra load of wood or gravel dumped into the back of a pickup truck, the company will get twice the value back in other ways.

A few miles north of Brookton, Chuck turns off Route 1 and onto Baskahegan's main east-west tote road, a smooth, wide, gravel thoroughfare where even a low-slung passenger car can cruise easily at 40 miles-per-hour. Chuck trundles along at ten, trying to keep the dust from billowing up and into the open windows of his small station wagon. The effort proves futile, and a fine film of yellow slowly appears on his sunglasses. He describes the history of each stand of trees we pass: how much Dead River cut, when they cut it, what he's trying to do now. On a bridge over Baskahegan Stream, with the black water sliding by below, he stops for a minute to talk about the early log drives. Then he puts the car back in gear, rumbles across the bridge, and pulls over in the sedges on the side of the road. He gets out and walks into the shady woods, ducking under branches of tightly packed spruce and birch trees.

Soon we pass a line of trees marked with thin pink surveying ribbon. Beyond the ribbon, many of the trees have been cut down, and the ground is covered with a thick mat of brush and branches. Here and there are piles of four or five freshly cut logs, neatly stacked according to species: fir, spruce, birch. Late morning sunlight sifts through the high branches of the remaining trees. The air is filled with the sticky sweet smell of balsam and the dull stink of diesel.

In the middle of the thinned part of the forest, a red machine about the size and shape of a farm tractor is moving slowly down a trail. The glass-walled cab is in the center of the machine, and we can see the operator sitting inside, his hands on levers and his feet on pedals. Mounted in front of the cab is a mechanical arm tipped with a heavy steel assembly. The assembly has two sets of opposing metal claws, two chain-covered wheels, a saw housing, and two stationary cutting edges that look like bent machete blades.

As we watch, the machine rolls forward on its huge tires and stops six feet away from a fifty-foot-tall balsam fir tree. The arm extends and the metal claws grasp the tree a foot off the ground. A circular saw slices out from under the claws and through the trunk of the tree. The arm lifts the shorn tree off its stump. The assembly on the end of the arm swivels until it is holding the tree four feet above, and parallel to, the ground. The operator pulls a lever, and the chain-covered wheels grab the tree and start turning, first one direction and then the other, driving the tree between the stationary cutting blades. Branches snap off against the blades at each pass, until only the naked trunk remains. The operator pulls a trigger, the circular saw whines through the trunk in a blur of sawdust, and a log exactly 100 inches long drops to the ground. The trunk rolls through the chain-covered wheels until the saw is again positioned at the 100-inch spot; another cut, another log. In ten seconds, five logs are lying on the ground. The arm drops the odd-sized end of the trunk and reaches out for another tree.

The Swedish-designed machine, called a Valmet Woodstar single grip processor, is one of only nine in Maine. While whole-tree harvesting machines have been common since the late 1970s, the Valmet processor is unique in that it fells, limbs, cuts logs, and sorts by grade and species right in the woods. The operator's instructions for this job are simple: Work only within the twelve-acre block taped off with ribbons. Cut all the smaller hardwood and balsam fir trees, and about half the spruce trees. Don't cut any pine trees. Stay on the trail. By the end of the day he will have processed hundreds of trees, leaving 30 cords of logs scattered along the trail and about a third of the original trees still standing. Tomorrow his partner will come by in a Valmet forwarder, a flatbed tractor mounted with a mechanical grapple, to pick up the logs and carry them out to the road.

Chuck had first considered using the Valmet "cut-to-length" logging system after seeing a demonstration in Sweden in 1988. At the time, all of Baskahegan's wood was being cut by hand crews (one or two men with chainsaws and a skidder) and by large whole-tree harvesting machines called feller-bunchers. These techniques worked fine on most of the company's land, but they presented real problems in high-quality spruce and pine stands, which Chuck was trying to cultivate with careful partial cuts. The trouble with the hand crews rose primarily because they were paid "piece rate" by the cord, and because Chuck required them to limb the felled trees and leave the branches in the woods to decompose. (It's easier and safer for the logger to skid whole trees out to the road and limb them there, or to use a mechanical de-limber.) When the weather was good and the terrain was smooth, the loggers could make an acceptable wage and still leave the residual forest in good shape. But in lousy weather or on steep or rocky ground, it was harder for the hand crews to earn any money. In their hurry, they would often end up felling or damaging the trees they were supposed to leave. Changing conditions had far less impact on the feller-bunchers, because the operators were paid by the hour, not by the cord. But the ponderous machines were not made for selective harvests, they couldn't limb the trees, and they didn't maneuver easily.

The Valmet system combined the best qualities of the two methods. As with a hand crew, the Valmet could move through tight spaces, pick out individual trees, cut the branches off, and buck the trunk into logs; the difference was, all these steps could be done much faster and in one step. As with the feller-bunchers, the Valmet could cut a consistently high volume, and the operator was paid by the hour, not by the cord. There were other advantages, too. Because the Valmet worked in tandem with a forwarder that carried the logs out of the woods instead of dragging them along the ground the way a skidder did, there was much less damage to the roots and trunks of remaining trees. Further, the mat of trimmed branches that built up in the forest sheltered the undergrowth from the weight of the machines.

In 1990, Chuck and Roger convinced one of their regular logging contractors, Colin Bartlett, to buy the first Valmet processor and forwarder in the United States. This was a big risk for Bartlett, because the two machines cost nearly $400,000. As it turned out, Bartlett made a good decision; the contracts he has secured with Baskahegan have been lucrative enough that he has since bought two more Valmet machines. Contractors for Seven Islands Land Company, Baxter State Park, and two other landowners are also regularly using cut-to-length systems, and their popularity is expected to grow.

Because Bartlett has to charge a high fee for the Valmets, Chuck uses them only in the Baskahegan Company's best stands, where the wood is valuable enough to justify the extra harvesting cost. Such high-quality wood amounts to only a quarter of the company's annual harvest, so most logging jobs are still done with feller-bunchers and hand crews. Chuck pays the hand crews a bonus to limb trees in the woods. The company has also experimented with horse logging.

"The techniques are all important," Chuck says. "If all we were doing was cutting with the Valmet, we would crash economically, because it would cost too much to get the wood out. I have to create a constant queue of forest stands, always mixing and matching what we need to sell with what we're doing in the woods." This mix of logging methods demands constant supervision. Chuck and Paul are in the woods every day, talking with the woodcutters, trying to be fair about spreading around the best wood and working conditions. If a woodcutter is getting frustrated, the foresters want to know about it.

Chuck watches the Valmet weave through the trees. Finally the operator, a young, wiry man named Byron Sanderson, who works for Colin Bartlett, throttles the machine down to idle and steps out of the cab. He pulls off his ear protectors and wipes his glasses as Chuck picks his way through the tangles of brush. The engine rumbles gently as they talk, waving their hands in a futile effort to shoo away black flies. The machine has been having braking problems, and Sanderson wants Chuck to know he was unimpressed with the mechanic Valmet sent to fix it. Chuck listens quietly, nodding, as Sanderson describes how he repaired the brake assembly himself. Now, Sanderson says, something is wrong with the alternator. The warning light's been blinking, and he's afraid it's going to go out on him. At last he shrugs and climbs back into the idling machine.

As we walk past the pink ribbons that outline the work area, the engine noise fades and the forest changes. Here smaller spruce and fir trees grow closely together in the cool shade. The ground is thick with ferns and with crumbly, rotting branches. We pick our way over fallen trees and mossy hummocks pushed up by the tangled roots and dotted with white trillium. Every few hundred yards the forest changes: light green waves of birch and beech on the low ridges, seams of red spruce following a wet seep, pockets of dead and dying fir trees crumpling into the forest floor. We emerge into sunlight, wade through a thicket of waist-high pine and poplar, cross a small stream, and are once again in a thinned stand of spruce and fir. Under a mat of decomposing brush, the ground is covered with tiny spruce seedlings.

Sanderson cut this stand with the Valmet three months ago, Chuck explains. He took out all the smaller hardwood trees and about half the spruce. The top branches of the remaining spruce trees, around 60 feet tall, provide a high canopy that filters light through to the forest floor. At Chuck's direction, Sanderson also left several 30- to 40-foot tall spruce, fir, and pine trees that are unevenly spaced between the taller spruce.

Chuck calls this a "modified shelterwood" cut, which is designed to encourage spruce regeneration by leaving some of the overstory intact. If Chuck had told Sanderson to cut all the trees, sun-loving hardwoods and raspberries would take over; this way, the remaining canopy will help the more shade-tolerant spruce seedlings establish dominance. Additionally, if he had not left some of the mid-sized trees, the stand would consist of just two generations: the tall trees far over our heads, and the tiny seedlings under our feet. By leaving the 30- and 40-foot trees, Chuck is maintaining a diversity of age, species, and size.

Chuck kneels, his knees squishing into the mossy soil, and lifts a small branch off the ground. In its shelter is an explosion of two-inch spruce seedlings, their tiny crowns a startling green in the lacy sunlight. He plucks one and sits back on his heels.

"This is the entire issue right here: regeneration. If you want to have some fun, get a bunch of foresters out here and ask them if these seedlings are established. Usually the academicians will say 'No', but the practicing foresters will say 'Dammit, they are so. Our living depends on it, and we've seen this stuff survive.' This has been the debate in Maine forever. If you were here with foresters for Great Northern, they would say, 'We can come in and clearcut this. We'll have to spray with herbicides, but this softwood is established, and we'll get good success.'"

He stands and holds his hand out chin-high, palm facing the ground.

"We want the new growth up to here before we let in any more sunlight. That will probably be in about fifteen years, and this smaller stuff will be completely established. Then we'll come back and take the overstory in one or two cuts. We won't take everything. We'll leave that spruce in the middle, and we'll leave the smaller pine trees over there to develop with the next stand. So we'll start to build some complexity and vertical diversity."

Chuck tries to define which areas of the forest to cut according to each stand's mix of species, their growing conditions, and their age. He might tell the loggers to cut out all the hardwood in one fifteen-acre stand, do a shelterwood cut in the next, and a partial thinning in the third. Some stands are left alone entirely. In others, especially thick stands of poplar and gray birch, he sometimes prescribes clearcuts in small patches or strips. His goal is to get the different sections of the forest "out of sync" with each other by cutting more lightly, but more often, than was done in the past. Chuck believes this approach is better than cyclical heavy harvests – not only for the land, because the resulting diversity contributes to forest health; but for the company, because Baskahegan always has a variety of wood available to sell to many different markets.

"If you talk to Great Northern about their forty- or fifty-year plan, they'd say in these two years we're going to put a road in and operate one township, then do the same with another township next year. What we're saying is, we are going to operate this stand, then that stand. It's a question of scale.

"This is a Maine version, but the concept of operating over your whole ownership every year is very much a European approach. Part of that is to have the discipline not to cut everything, even though a stand could benefit from a partial cut, because we don't want to lose the overall effect."

All large timberland owners have the same basic goal. They want to replace cut trees with new trees of an equal or greater economic value (in Maine this often means pine and red spruce), and they want to do it fast. The Baskahegan Company believes the best way to do this is to invest before and during the harvest with good roads, careful logging methods, close forester supervision, and by moving around throughout the ownership. By spreading around the cut and leaving some cover, the forest will regenerate naturally with virtually no assistance. Many other landowners, especially pulp and paper companies with mills to feed, look at it the other way around; they invest after the harvest to speed regeneration of the desired species. They build a road into an area and cut the wood as inexpensively as they can, often in large clearcuts or strip cuts where all the trees are harvested in patches or rows. When the hardwood "weeds" that thrive in the sudden sunlight and warmer soils start competing with the more valuable spruce, the companies often knock them back with herbicides that affect only deciduous species. If the site really wants to grow hardwood and the mill needs softwood, the company plants spruce or pine, sprays herbicide to kill the hardwoods, and thins the planted trees with hand-operated brush saws. Then they come back forty or fifty years later and do it all over again.

"If you clearcut, you are starting truly at ground zero," Chuck says. "But when you release regeneration this tall, the overstory is still growing and developing. There is great efficiency in that." He holds up his little seedling. "It takes a lot of money to plant red spruce trees like this well. Here, they're just handed to us."

Chuck generally resists comparing Baskahegan's forest operations to those of other timber companies. Maine, he says, is not one forest. From region to region, soils are different. Stability in the stands is different. Regeneration is different. Inter-species dynamics are different. Landowners, too, are different. Unlike Great Northern-Bowater and other paper companies, the Baskahegan Company does not own a mill that requires a certain amount of wood every week, the company's five shareholders don't demand a high quarterly dividend, and the Millikens' wealth largely shields them from the timber industry's ups and downs; as Chuck puts it, Baskahegan does not set the Milliken family's standard of living. As a result, Chuck works under very different constraints than do many other foresters. And while he's proud of what he does, he calls it "an embarrassment" compared to what he has seen in the meticulously managed forests of Germany and Sweden.

"What is good forestry? What is bad forestry? It's just not that simple," Chuck says. "It's rare to say something is a complete success or a complete failure. What pulp and paper outfits look for is to grow fiber. What we look for is to grow value. It just happens that fulfilling our mission is a less grossly manipulative kind of management."

Chuck drops the tiny seedling on the ground. With his foot he nudges the rotting branch back into place over the little nursery. We walk deeper into the forest, the high branches above our heads swishing in the spring breeze.

8

Chasing Chances

In 1979, a young forester named Brian Souers was working for International Paper Corporation in Lincoln, forty miles west of Baskahegan Lake. That year, Brian bought some land and a horse and started dubbing around in the woods on weekends, cutting firewood and selling an occasional load of logs. He liked being outdoors and in 1980 decided to try logging full time as an independent contractor. He left his job with International Paper (IP), bought another horse, and landed a contract doing selective thinning on the company's land. Brian would cut trails into the woods in a herringbone pattern and fell the trees so their butts faced the same direction. Then he would back the horse up, hitch the animal to the logs, and coax them out to the road.

This kind of logging was called timber stand improvement, or "TSI." The objective was to remove some of the trees, especially poorly formed or crowded ones, to increase light and nutrients to the remaining stand. The trees that were left would then grow faster, becoming more valuable to International Paper. Brian could afford to do a careful job, because IP was paying him not by the volume of wood he cut, but by the acre. After several months he incorporated under the name Treeline, Inc., and bought a German-made Holder skidder – a small, highly maneuverable machine specially designed for TSI. After a year he hired a laborer and picked up another Holder. The money was modest but steady, and Brian and his wife Denise started adding on to the old fishing camp they had bought as a home on Mattanawcook Pond in Lincoln. Their first son was born in 1981.

93

Then IP ended the thinning program. At the time, wood prices in Maine were dropping, the nation was in recession, spruce budworm was threatening millions of acres of timber, and the company could no longer justify the cost of the TSI.

With the major landowners putting more emphasis on volume, Brian secured a conventional contract to cut 4-foot long pulpwood logs from IP's land. He started working longer and longer days.

"That was a tough change," Brian says. "I was twenty-four years old and had borrowed sixty thousand dollars at eighteen-and-a-half percent interest for those German skidders." Now he had to use the small machines in strip cuts and clearcuts, which were much more extensive than the TSI jobs for which the Holders were designed. His competitors, equipped with much larger skidders and whole-tree harvesting machines, could cut several times as much wood in a day. Every month Brian fell further behind, until finally he was faced with making a choice: He could quit logging; he could look for small private landowners who would hire him to do thinning and low-impact selective harvesting; or he could take the plunge into larger equipment and mechanization, which were fast becoming the only viable means of competing for commercial logging contracts.

For Brian, the first two options were out. He had found something he really enjoyed, and he wanted to be his own boss. He was not quitting. Thinning jobs were hard to find in central and northern Maine, however, where large commercial timberland owners controlled three-quarters of the forest. The few small landowners who wanted light harvests done tended to hire veteran woodcutters. That left the third option, and Brian took it.

Working long hours and making use of his contacts at International Paper, Brian started to expand his company. For eight years running, he invested a minimum of $100,000 a year in new equipment. He put together a fleet of big skidders and hired experienced woodcutters to run them; one year, five Treeline loggers brought in the staggering volume of 3,500 cords apiece. Brian bought two whole-tree feller-bunchers, a mechanical delimber, a portable saw called a slasher that cuts trees into logs, and several log trucks. In 1986 he hired another forester, Dick Slike, to handle the books. Dick and his wife Katy bought an old camp just down the road from Brian and Denise's place, and the two ran the business out of the Souers' home.

The whole-tree harvesting machines could cut wood for several dollars less per cord than a man with a chainsaw and skidder could. But expenses, most notably loan payments on the equipment, maintenance, insurance, and repair, were rising, and Brian found he had to cut more and more wood to meet his debt and keep Treeline in business. Most of the jobs were service contracts on land owned by International Paper and by log broker H.C. Haynes. Once in a while, Brian would get a stumpage contract and sell the wood directly to the lumber and paper mills. He also bought several pieces of land, cut and sold the better quality wood, and then subdivided and sold the land. By the late 1980s, Treeline was cutting 45,000 cords a year. Because of high expenses, however, a third of that volume earned no profit for the company.

Other contractors were on the same treadmill, propelled ever faster by the mills' policy of paying an extra "incentive" rate to high-volume producers. The message was clear: grow or die. The fierce competition kept wood supply to the mills high and prices low as the annual harvest in Maine swelled from just under six million cords in 1979 to well over seven million in 1988. Yet employment declined as mechanization reduced the number of jobs in the woods.

"The logging business is like a snowball," Brian says. "You start it small and start rolling it along and it generates momentum. It just kind of happens. You end up doing the things you *have* to do, rather than the things you *want* to do. As the margin gets tighter, what do people try to do? They cut more wood to make up for it. And what does that do? It makes the margin even tighter by driving supply up and prices down."

In 1990, Souers and Slike realized that Treeline could clear the same amount of profit by downsizing – cutting smaller volumes of wood and lowering overhead. They sold several machines and hired woodcutters who owned their own skidders. The company's annual cut dropped by around 10,000 cords, with yearly equipment costs reduced to around $20,000. Brian started looking for other ways to make money. He built a service garage on Route 2 just outside of Lincoln and bought a large piece of land for a subdivision on Lake Mattanawcook.

In 1992, Treeline lost the Haynes contract and signed on with Baskahegan Company. By 1994, Treeline was back up to an annual cut of around 45,000 cords, mostly coming from forests owned by International Paper, Baskahegan, and Boise-Cascade, another paper company. Now Treeline owns an excavator, a bulldozer, a road grader, a delimber, three skidders, two slashers, and three log trucks. The company cuts about 75 percent of its wood with hand crews and subcontracts with a feller-buncher crew for the rest. In all, the work keeps ten loggers, three equipment operators, and three truck drivers busy. They deliver some twenty-six different combinations of species, grades, and log lengths to twenty different mills. Half of this wood is used to make manufactured lumber and studs, and another 20 percent goes as higher-grade sawlogs. The rest ends up as pulp or, on the other end, as high-quality hardwood veneer.

Some of the jobs Treeline accepts make money; some don't. The profit is still so narrow that the company often has to take marginal jobs just to stay in business until better opportunities come along.

They call it chasing chances.

"There's just not that many good stands out there," says Dick Slike, Treeline's money man. "It can be two-and-a-half years now between decent chances." He pauses and shifts his weight. "It can be longer."

96

Dick is leaning against the doorjamb of a cluttered office at Treeline's cinder block garage in Lincoln. Tim Shorey, the company's logging foreman, is sitting on a desk. The two look as if they've swapped roles: Dick, stocky and powerful with shoulder-length black hair and a thick beard, looks more like a lumberjack than Tim, who is leaner and clean-cut. They both wear work boots, jeans, and tan shirts emblazoned over the left pocket with the company name as well as their own. The room's particle-board walls are papered with safety posters and government NOTICE TO EMPLOYEES bulletins. From the garage comes an occasional shout and the clang of metal on metal.

The small room is crowded; both men are over six feet tall, and they had not planned to stop and talk here. We had been on our way out with Tim when we ran into Dick. As invariably happens in conversations about logging in Maine, money had crept into the discussion, and Dick had offered a few thoughts. That was a half hour ago.

"We cut gobs and gobs of wood that we make no money on," Dick says. "There's a lot of gross to make. The trick is getting the net. We bid for stumpage deals where we're going to make a dollar a cord, and we lose them. How is that possible? Who can cut wood for less than a dollar a cord?"

Nobody, Dick says, answering his own question, unless they are cheating. All contractors who follow the rules have similar fixed costs. Workers' compensation insurance alone costs Treeline $42.67 for every $100 the company pays in wages. On top of that comes health insurance, Social Security, safety training, fuel for company trucks and machines, maintenance, taxes, and mortgages on the garage and equipment. Dick knows the company loses jobs to contractors who aren't covering some of these costs because they don't buy workers' compensation insurance, or because they pay their loggers under the table.

Down time is also expensive. Treeline gets its best chances to cut quality wood in the winter, when wet areas that grow good spruce are frozen and the heavy equipment does less damage to the ground. Problem is, that's also the most difficult time for trucking, and Treeline doesn't get paid until the logs and chips arrive at the mill. Hundreds or even thousands of cords can pile up in the woods if the roads get soggy or a truck breaks down; meanwhile, Dick has to keep writing paychecks. These cash-flow problems are even worse for many other contractors. Treeline keeps repairs in-house, and the company's good safety record holds their workers compensation premium as much as 25 percent below some of their competitors.

"Once in a while we run into a deal in the woods where Treeline makes money, and the woodcutter makes money, and the mill makes money, and it's a win-win-win situation all the way around," Slike says. "But time and time again there are two or three people in this chain who are losing. There's all this lose-lose-lose stuff."

Most often, the biggest losers are the people doing the most dangerous work: the woodcutters. Like most contractors, Treeline pays its loggers by piece-rate: the more they cut, the more they earn. Treeline's rate runs between $4 and $5 per cord, depending on the species of tree, and depending on whether the logger can drag out whole trees or has to spend time cutting off limbs in the woods. To keep its payroll and therefore its insurance costs down, the company also pays equipment rental fees totaling around $10 per cord to loggers who use their own skidders and chainsaws. A steady woodcutter who brings in 50 cords a week will gross about $700 in wages and equipment rental fees. Out of that comes fuel for the skidder (around $100 a week), payments on the skidder, gas and oil for the chainsaw, and equipment maintenance. Logging machines are expensive. An average-sized cable skidder costs around $50,000 and can run as much as $65,000; a set of tires for a skidder goes for about $5,000. While these expenses rise continually, the piece-rate hasn't changed in eight years. It's all many loggers can do, Dick says, just to clear minimum wage. If you can't increase the amount of money you clear from every cord, you have to cut more cords.

"Corporate America has to make its money," Dick says. "Then comes the unionized workers in the paper mill, then the logging contractor." By the time the fixed costs are paid, hourly salaries for other employees covered, and Brian Souers takes his share, there is little left over for the individual loggers. "A woodcutter's out there taking his life in his hands day in and day out. And he's getting paid, like, minimum wage to do it," Dick says.

The phone rings and someone down the hall picks it up. Dick had ignored a few other calls during our conversation, but this one is from Brian. He says he'd better take it, and disappears into his office with a wave.

"Well," says Tim, standing up. "To work."

Pulling out of the driveway in an aging green Ford pickup truck, Tim lays out his day's itinerary. Treeline recently signed a ten-week contract with International Paper to cut 7,000 cords of wood from Township 3 Range 1, a 23,000-acre square of virtually uninhabited timberland just east of Lincoln. Tim needs to check in on the hand crews, write down how much wood they have cut, refuel a skidder, and make sure that the men running the delimber and slasher have enough wood to process. Treeline will barely break even on this contract as long as everything runs smoothly, and it is Tim's job to make sure that happens. He switches on the two-way radio and drives east over the Penobscot River. In the middle of Lincoln's busy three-block downtown, he bears right on Route 6 toward Township 3 Range 1.

When northern and central Maine was originally surveyed in the nineteenth century, government agents divided the region into a grid of square townships, each measuring 36 square miles, or about 23,000 acres. Ten of these townships constituted a range; Township 3 Range 1, or "T3R1," was the third township in the first range west of Maine's border with New Brunswick. As settlers purchased land and formed local governments, they replaced the township numbers with formal names, such as Lincoln, Topsfield, Brookton, and Kossuth. Settlement passed many areas by, however, and the government sold scores of politically unorganized townships to timber companies and entrepreneurs. Today, International Paper owns all but a few hundred acres of T3R1. The only residences are hunting camps.

International Paper, a multinational corporation that ranked 31st on the Fortune 500 in 1994, owns a paper mill in Jay, Maine, a factory that makes insulation boards in Lisbon Falls, and another that makes containers in Presque Isle. IP Timberlands, a stand-alone business within the larger corporation, owns 970,000 acres in Maine, another 300,000 acres elsewhere in New England and New York, and more than 6 million acres nationwide. Working out of several offices, including one in Lincoln, IP foresters lay out roads, write management plans, and hire logging contractors to harvest and truck the wood to buyers across the state. Treeline has been the principal contractor on T3R1 since 1985. Brian's crews built most of the roads there, and they spend a lot of time and money maintaining them. Treeline's woodcutters cover about a thousand acres of T3R1 every year, harvesting some 15,000 cords in a variety of methods. Only around 10 percent of this wood ends up in International Paper's own facilities.

Deep into the township, Tim halts the truck in the middle of a gravel road. On our left is a thick forest of hemlock and poplar and beech trees, most between 6 and 12 inches in diameter. On the right is a completely treeless, 35-acre rectangle, shorn of all vegetation and scraped smooth, like an oversized soccer field awaiting new grass and white lines. On three of its four sides, the clearcut ends abruptly in straight lines of trees. On the fourth side, along the road where Tim has stopped the truck, lies a house-sized pile of logs. Their raw ends are moist and pitchy and tagged with numbers in fluorescent orange paint: 6, 7, 6, 8, 6 , 6, 5, 5, 6, 5, 5.

Tim steps out of the truck and looks at the empty land. IP, he says, prefers the term "stand conversion" to "clearcut," because the company's goal is to convert a forest that was growing mostly hardwoods and a bit of hemlock to one that will grow a crop of pure spruce.

"This was probably 90 percent hardwood and 10 percent softwood, and they want to convert it to all softwood," Tim says. "This was a good chance for a woodcutter, because all he had to do is cut it all down and bring it to the road. That's where the woodcutter makes money." After the logger cut all the trees and hauled them to the roadside, Tim explains, Treeline's processing crew came in with the mechanical delimber and the slasher and turned the trees into logs. The tops and branches, along with whole, smaller trees, were fed into a giant chipper, and the chips were blown into a tractor trailer to be hauled to the mill. The logs will be picked up later this week. Next summer, laborers working for IP will come in and plant spruce seedlings. The following year, they will spray herbicide to kill the hardwood saplings that by then will be threatening to choke out the spruce transplants.

From the woods on the far edge of the clearcut comes the heavy droning sound of a diesel engine. We walk on the road along the clearing until we reach the line of trees, then follow the sound into the forest. Two hundred yards off the road, in a stand of poplar, hemlock, and fir trees, Treeline woodcutter Mike Dube is unreeling a thick steel cable from a winch mounted on the back of a yellow John Deere skidder. We stand and watch him work.

Dube has already felled several poplar (or "popple") trees. When he has the cable pulled out to one of the downed trees, he hurries back to the skidder for a chain. Returning, he wraps the chain around the butt of the tree, hooks the chain to a sliding cuff on the cable, scurries back to the skidder, and engages the winch. The revolving drum pulls the cable tight, dragging the fallen tree into line behind the skidder. Dube pulls the cable back out and attaches it to another tree. While this one is being reeled in, a limb catches on a stump. Dube climbs into the cab of the skidder and tries to drive the machine forward to jerk the log free, but the limb holds fast against the stump, and the front wheels of the skidder lift off the ground. He jumps down, releases the tension on the winch, reaches behind the seat of the skidder, and pulls out a chainsaw. He follows the cable through the brush and branches back to the poplar, slices off the hung-up limb, and returns to the skidder. He engages the winch again, and the cable draws the tree into position next to the first one.

It takes Dube fifteen minutes to winch five trees into a bunch directly behind the skidder. Then he holds the winch lever down until the cable lifts the butt ends of the trees several feet off the ground. He climbs back into the cab and drives forward. He tries to maneuver around a 20-foot standing poplar, but every time he turns the skidder, the load of trees dragging behind the machine gets caught on a stump. Finally he gives up, roars straight ahead, and the small tree disappears under the skidder. With the huge rubber tires churning the soft ground, Dube drives the skidder out of the woods and across the adjacent clearcut. He stops next to the pile of logs alongside the road, not far from Tim's truck.

We follow him on foot across the clearcut.

By the time we get to the road, Dube has released his load of trees and is standing awkwardly next to his skidder. He looks distinctly apprehensive, like a kid waiting to see the principal. When Tim makes introductions, Dube's face sags in relief. "Jesus Christ," he says, blushing, reaching out to shake hands. "You guys scared the shit out of me. I thought for sure you were OSHA." Dube rarely sees strangers at his job sites, which are usually many miles from the nearest paved road. He figured the sight of his boss showing up with two guys carrying notebooks and cameras could only mean one thing: the Occupational Safety and Health Administration. We know that loggers dread visits from OSHA inspectors, but Mike's fear seems to be out of proportion. All of Treeline's woodcutters go through safety training, and Mike is wearing the obligatory hard hat, face screen, gloves, steel-toed boots, and safety trousers made of Kevlar (the material used in bullet-proof vests).

He had handled his saw well, and the skidder appears to be in perfect operating condition.

But an OSHA inspector, Dube says, could have found plenty to complain about, plenty to fine. Dube's gloves aren't of regulation thickness. His cable is nearing its recommended maximum number of hours. A regulator would have objected to the way he pushed over that small poplar. Most offensive of all, he had left a few standing dead trees in his work area. OSHA regulations require loggers to cut down every dead "stub" more than 2 inches wide and 10 feet tall within two tree lengths of their work site. Since the budworm outbreak, few forests in Maine are free of standing dead fir trees, and many areas are completely choked with them. Since Dube was cutting 50-foot poplar trees, technically he first should have felled every dead tree within 100 feet of where he was working at any given time. He had dropped most of them, but if he followed the law to the letter, he would never make any money — nobody pays him to cut dead stubs; they just tell him he has to. If he doesn't, and if he gets caught, Treeline is fined $1,000 for each stub. And OSHA looks. Thirty-four loggers were killed on the job in Maine from 1978 to 1993, including several who were crushed under falling dead trees. Treeline has recently gone through a pre-inspection audit, and unannounced field checks are imminent. That, Dube says, is why we scared him so badly. He keeps saying it every time the conversation lags, and he repeats it before climbing back into his skidder.

"Jesus," he says, shaking hands with us again. "You guys sure scared me."

"This is not a good chance for Mike," Tim says, as he watches the skidder disappear back into the forest. "He can make his wage, but he's not making any money. But right now there's nothing else to cut. It's one of them deals that you've got to put up with for a couple, three weeks. And then you get back into better wood."

Treeline pays its woodcutters on a "butt scale," which is based on the diameter of a tree on its thick end. If all the trees a logger has cut and piled are 6 inches across, he gets paid the same for the 50-foot tall trees as he does for the 30-footers, even though the taller ones are more difficult to fell and skid out of the woods. The landowner, of course, gets paid more for the taller trees when they arrive at the mill, because they contain more logs. Thus neither the woodcutter's pay nor the difficulty of his labor is reflected by the value of the trees he is cutting. Loggers paid on the butt scale particularly dislike thick stands of fast-growing hardwoods such as poplars, because they are usually full of skinny, tall trees.

Tim knows exactly how much money Dube is making, because part of Tim's job as logging supervisor is to tally what the loggers cut. He comes by twice a week, measures the diameter of the logs, paints the measurement on the butt, and reads the tally into a hand-held tape recorder. A secretary listens to the tape and fills out tally sheets so Dick Slike can write out the logger's paycheck.

"Tim just talks into the tape," Dick had explained earlier. "'Mike's cut: hemlock . . . five, hemlock . . . six, spruce . . . four.' Although these days it's more like, 'popple five, popple five, popple five, popple five, popple five, popple five.'"

Driving down the gravel road, Tim lifts a folder off the dashboard and extracts a letter-sized, photocopied topographic map marked with International Paper's arrow-shaped corporate logo. The map is dense with squiggles and polygons, hashmarks and large capital letters. It takes a moment to figure out where we are and what we're looking at. Alternating his attention between the road and the map, Tim explains the legend: hashmarks for site conversions, A for release cuts, B for strip cuts, C for basal area cuts.

As the landscape around us rolls past, it's not hard to follow the map's symbols. Here is a "B: Strip cut," where 20-foot-wide treeless lanes gouge a quarter mile into the forest, separated by 40-foot strips of uncut woods. These alternating strips of brown and green tick past for half a mile and then stop abruptly at "A: Release," where the forest has been reduced to a scattered few spruce and birch trees, like birthday candles on a five-year-old kid's cake, frosted with a tangle of undergrowth and shattered branches. Then we're into "C: basal area cut," where the loggers were told, essentially, to take half the trees and leave the rest standing. Just when this more parklike look starts to become soothing, the forest changes to "Hashmark: site conversion," where the bald earth waits to be plugged with seedlings. Large areas have also been cut by "diameter limit," where all of the trees thicker than 4 or 5 inches have been removed. Every few miles we pass a plantation of shoulder-high red pine or Japanese larch trees, lined up in straight queues that disappear over the tops of ridges.

Only the site conversions, about a fifth of the total harvest in T3R1, are replanted; IP counts on the rest of the areas coming back through natural regeneration. Strip cuts are supposed to reseed themselves from the band of uncut trees on either side, which also provide shade to keep the soil cool; release cuts are done only where there are seedlings of the desired species already established, and the loggers are told to leave a few seed trees; and basal area cuts retain enough residual trees to mimic natural succession. And yet, despite the diversity of these "prescriptions," as foresters call them, there is a numbing sameness in the land. The shoulders on both sides of the road are cleared 30 feet back to accommodate countless piles of logs and brush. In the stands where some trees were left, many now lean over or have fallen into piles like pick-up sticks, blown down once the protective forest around them was gone. Virtually no large trees, alive or dead, are left in any of the cut areas. Live large trees were cut and hauled to the mill; the dead ones, which would have helped feed and shelter many species of wildlife, have fallen to OSHA's rigorous regulations. Fragments of wood and bark are everywhere.

In the text of its recreationists' map of Maine, IP Timberlands declares that the company uses "a variety of forestry techniques to work with nature. The result is a healthier, more vigorous and more valuable forest. Recognizing the environmental and visual impact of harvesting activities, we limit the size of harvest operations and intersperse them among stands of different ages. In addition, harvest operations are usually irregular in shape to maximize the benefits for wildlife . . . Our foresters have developed and implemented Environmental Guidelines for use in all IP Timberlands forests. Many of these guidelines go beyond measures stipulated by law . . . Our foresters require that all contractors operating on IP Timberlands adhere to these guidelines. On-site inspections are conducted to ensure compliance."

In broad terms, this text accurately describes what happens on T3R1. The company does use many forestry techniques to manage and harvest trees, and certain areas undoubtedly grow back more vigorously as a result. Foresters from International Paper use maps and ribbons to define the prescription for each block. They check in regularly with Tim and Brian to find out how the harvests are going, and they monitor mill receipts to make sure Treeline is sending in the right volumes of wood. Paper company foresters are trained to understand the impact of logging on soils, wildlife, scenery, and water quality. Many are active in a range of professional and scientific organizations, and they work closely with state wildlife agencies.

According to many loggers, however, paper company foresters make few day-to-day decisions in the woods. A logger can work for a month without ever seeing an official from the company that owns the land. In that time, the logger might remove 200 cords of wood; a feller-buncher will cut five times that. Even when foresters do show up, they are forbidden for insurance reasons to give instructions directly to the woodcutters. Direct contact suggests an employer-employee relationship, which the paper companies avoid lest they become liable for injuries, workers compensation, or payroll taxes. For much of the harvests, of course, on-the-ground forestry instructions are somewhat irrelevant: In site conversions, the loggers cut everything down to the dirt. In strip cuts they do the same, but in rows. In release cuts they leave a few trees and try not to crush the little seedlings. In diameter limit cuts, they cut every tree that is wider than a certain diameter.

Across from a sprawling plantation of feathery, 8-foot larch trees, Tim turns left on to a deeply rutted road and stops. Three men in tan Treeline work shirts and jeans are huddled over the saw housing of a slasher, a portable saw that Treeline hauls to logging jobs behind a truck.

"I hate Mondays," says Rick Irish, the slasher operator, as Tim walks up. A flywheel in the motor has shifted slightly and sheared off several bolts. Tim peers into the machine and shakes his head. He takes the parts manual out of Irish's hand and pulls a tiny notebook from his own shirt pocket. After carefully comparing the actual broken parts to the illustrations in the manual, he copies down several numbers and goes to the truck to radio the garage. If the parts aren't in stock, the machine could be down for two days. If the garage has them, Tim will bring them back tomorrow and fix the saw.

The slasher is a simple machine. It has a long steel carriage with four sets of arms sticking up at forty-five-degree angles to cradle the logs. In the middle of the carriage is a large circular blade and motor. The truck that pulls the slasher is equipped with an extendable grapple claw called a cherry picker. Sitting in a cab, Irish picks up long logs with the grapple, lays them into the carriage, and runs the blade through to cut logs of 4, 8, or 16 feet, depending on what the contract calls for. Once the logs are cut, he uses the grapple to pile them by species, size, and grade. A slasher generally works in tandem with a mechanical delimber. Treeline's delimber, operated by Arthur Ripley, is parked just down the road from the slasher, in front of an immense tangle of downed trees that still have their tops and branches intact. The delimber has a 55-foot, two-section, telescoping boom. On the end of the outermost section is a grapple and a doughnut-shaped cutting head. A second grapple is mounted in the middle of the other section.

The end grapple picks a tree out of the pile and swings the butt end up to the middle grapple, which holds it tight as the sections of the boom telescope in and out, driving the tree through the cutting head until the branches are sheared off. When the tree has been cleaned of its branches, Arthur swings the arm around and deposits the trunk into a pile where Rick can reach it with his cherry picker. When the two machines are running together, they look like praying mantises, weaving and bobbing in an intricate dance. They start with the more valuable softwood logs, then the softwood pulp, and then the hardwood pulp. Arthur is usually ahead, delimbing up to 100 cords a day, while Rick saws and sorts around 60 cords. That's how it should work, anyway. When the slasher broke down, they had been working their way through a pile of trees that Bill Emery, one of Treeline's best woodcutters, had cut down and dragged out of the woods with his skidder.

"Someone's going to have to go to town and get me some pills for this depression I'm in," Rick says when Tim returns from his pickup truck. "You brought me over from that big orchard in Martin and into this. I was just getting into those big rock maples when you moved me. That was kind of depressing."

Bill Emery seems amused at Rick's griping. Wood size and mechanical hassles are of no consequence to Rick and Arthur, because they are paid by the hour, and Treeline owns the equipment that they operate. If it had been Bill's skidder that had failed, he would be losing twice: He'd have to pay for repairs, and he wouldn't be earning any money as long as he was down. On the other hand, on good weeks Bill will earn more than the other two will. He's into a good chance in this site, release cuts in fat hemlock and mixed hardwood. He has no selection decisions, doesn't have to limb in the woods, and can use his large grapple skidder to haul several whole trees out of the woods at once. The only thing slowing him down is a new IP rule: After Arthur has delimbed all the trees, Bill has to use his skidder to push the piles of branches back into the woods, rather than leaving them along the road. But the twenty minutes spent every day pushing brush is nothing, he says, when weighed against the time it would take him to limb trees in the woods with a chainsaw. "The limbs stay on your body better, too," Rick quips.

The men go back to work: Tim to the broken saw, Bill to his skidder, and Arthur to his delimber. Rick is idle. He walks over to a pile of wood and peers closely at a large maple log. The men are always on the lookout for "money trees," birds'-eye and curly-maple logs that are prized for fine woodworking. Buyers will sometimes show up in the woods and dig into a log to see if they can find the distinctive patterns. "It has a ripple, little bumps," Rick says, raising his voice so he can be heard above the steady roar and sporadic crash of Arthur's delimber ripping the branches off a stubborn hemlock. "When you plane it, it still looks like it's rough, but it's smooth. It pays big money. *Big money.* I sawed one log last year, one 12-foot log, worth *eighteen hundred dollars.* Not to us, but to the people I sawed it for." Finding money trees doesn't happen very often, and this log turns out to be ordinary hard maple.

Rick leans against the pile and folds his arms across his chest. "This used to be the pine tree state," he says. "It's more like the alder state now. Or the larch state. It just doesn't have the value for the woodcutters anymore. The landowners are just squeezing it for all it's worth."

He looks up the road at the plantation of evenly spaced larch trees.

"Now wouldn't you be proud to own that?" he says wryly. "I know I would. It's a shame to walk in the woods and see something like that. I just don't like to see it. But I have to feed my kids, and that's what buys my biscuits at the end of the week. I make a living at this. But I don't have to like it."

Only Baskahegan, Rick says, seems to have a different view than the other companies that Treeline works for.

"I'm not saying they're right or wrong, they're just different. But they do have wood left when they're done. When nobody else has any spruce left, Baskahegan will still have some."

Brian Souers is sitting on the porch railing of a brand new log-cabin home on the shore of Mattanawcook Pond. The door and window frames gleam with fresh green paint; scraps of wood litter the ground. The house is set back some distance from the lake, which is visible through a screen of large pine trees. The sound of a hammer comes from the next lot over, where two carpenters are hanging siding.

Brian is showing us his most ambitious attempt yet to diversify Treeline's revenue: a subdivision called "White Point Estates: A Lakeshore Community." He bought the 530-acre property in two parcels in 1990, intending to harvest the mature white pines. But the trees seemed to be more valuable standing on the point than going through a mill, and Brian decided to develop one of the two tracts. He assigned Treeline loggers and heavy equipment operators to cut in a road and create thirty-three lots, two or three acres each. The lots are well-spaced, screened from one another and from the lake, and shaded by tall pines. The project consumed much of Brian's attention for several years as he experienced what he calls the "pleasure and education" of applying for permits from the U.S. Army Corps of Engineers, the Maine Department of Environmental Protection, Maine Inland Fisheries and Wildlife, and the local planning board. Now the lots are finally starting to sell. He is proud that most of the buyers are local people.

Of this venture, Brian says, "People invest their time, energy, and money into forest products for one reason: profit. When the risk is too great and the profit is not there, you can sit and butt your head against the wall, or you can begin to shift your resources and energies to something else. That's the checks and balances in the system. As people move out of the industry, that will actually help get its health back."

White Point is the largest development project that Brian has attempted, and it has created friction in the company by diverting his assets and time. Employees complain that the boss is no longer as focused on finding good wood to cut, that long overdue raises will never materialize, that the woodcutters are paying for a poorly timed real estate deal. But the griping is muted. Brian Souers has saved more than one logger from bankruptcy, and Treeline's benefits package is more generous than those offered by most contractors. As Dick Slike points out, if White Point had not been developed, the only way to generate a return from that land would have been to strip all the big pine trees. If that had happened, he says, the point would now be choked with pin cherry and gray birch trees. Brian and his family still live on the lake, and he wanted to do something he could live with.

Nevertheless, Brian knows that White Point was a tremendous risk. He started the project just as the real estate market was waning in 1989, and he's lucky that the lots are selling as well as they are. For now he's not investing in any new projects. He's keeping his options open, though, looking for different kinds of chances. He knows too well the risks of relying solely on logging.

"There are so many increased costs that we really don't have control over," he says. "Most contractors are walking tightropes. It doesn't take too much of a goof-up. You buy the wrong machine at the wrong time, get your debt service out of whack, and then have a little hiccup – say a particular mill stops buying wood for three months – and you're out.

"I'm not a logger logger. I'm a business logger. It's not unique to be a business logger, but it's not as common as just being a logger logger."

Treeline logger Andy Ward pushes his plate aside and picks up a notebook. "Okay," he says, "let's talk about skidding distances." With a pencil he sketches a rectangle 6 inches high and 4 across. He says, "This is the block I'm working in now." He draws a quarter-inch vertical line from the base of the rectangle up toward its center. "And this is the road. It's forty-two chains to the back." A chain is 66 feet. "So I have to waste my time and fuel going back and forth more than half a mile from the road to the back of the lot. The road system should be like this." He draws a broad loop within the rectangle, like a lasso lying in a shoe box. "Then, from every point, it would be no more than eighteen chains to the road."

Dick and Katy Slike are sitting at their kitchen table at twilight, listening to Andy, a stout, compact man with a round face and tight curls of dark hair. The Slikes' six-year-old son, Levi, is rolling a toy truck around on the linoleum floor. The table is covered with the remains of dinner: steak and baked potatoes that Dick had cooked on an open fire in the backyard. Through a large picture window we can see the broad expanse of Mattanawcook Pond, its late winter ice shrinking back from the shores.

Katy looks at the sketch. If Andy's road layout makes more sense, she asks, why doesn't the paper company do it that way? Dick explains that it's mostly a matter of cost: Gravel and equipment is expensive, and International Paper wants to keep as much land as possible in wood production. He points out that wildlife ecologists, too, are always urging the large landowners to avoid fragmenting habitat.

Andy shakes his head. "They're killing themselves doing it the way they are now," he says. He believes that the paper companies waste more money and forestland by clearing unnecessarily wide shoulders for the existing roads than they would if they laid out more modest lanes that reached further into the logging areas. To Andy, the frustration caused by the current road system, which forces him to spend extra unpaid hours in his skidder, is more than a problem of time and money; he sees it as part of a troubling trend away from thoughtful planning in forestry. Another example is strip cuts. Andy says that the paper company foresters used to come out to the woods with a roll of surveying ribbon and carefully mark each strip that was to be cut, and each strip that was to be left. Now, he says, they just mark the edges of the entire block. Then they tell Treeline how wide each strip should be within that area, leaving the actual measurements to the loggers.

"I guess the ribbon was too expensive," Andy says with a tight smile. He knows that it's a question of man-hours, not ribbon, and that bugs him even more, because he's the man who has to spend the hours. Ironically, such extra work is even more troublesome in partial cuts. While Andy favors partial cutting in principle – he thinks it's the best way to ensure there will be wood left for the future – in practice he earns more money in strips and clearcuts where he doesn't have to worry about banging up the residual stand. In some partial cuts, he also has to spend time figuring out how many trees he is supposed to take out. He does this by calculating "basal area," the combined horizontal surface area of the tree trunks at 4½ feet off the ground. Traditionally done by foresters, estimating basal area is a time-consuming and technical job requiring a glass prism and a calculator. "They expect me to do all their legwork, they expect me to do it right, and I get nothing for it," Andy complains. "I'd like to see the foresters do more work. Way more."

Andy has been with Treeline since 1989, when he moved east after logging in Wisconsin for a decade. He likes being in the woods, takes pride in his work, and hopes to do it for many years. In order to squeeze out a livable wage, he tries to keep his controllable costs down. He runs a used skidder, drives a used pickup truck, and keeps them both well-maintained. He is careful to separate his wages from the rental fee Treeline pays him for using his own equipment, to avoid paying unnecessary income tax. When uncontrollable events threaten his livelihood, such as the time someone attempted to torch his skidder on Christmas Eve in 1992, he tries not to panic. To keep from making potentially tragic mistakes, Andy sets an annual income goal rather than a weekly target.

"If you have to go to work on the rainy days just to make your payments, cutting wood for a living can get pretty miserable," he says. "And if things aren't going right, it can become disastrous. You got these guys out there with two-hundred-and-forty-thousand-dollar whole-tree harvesters with seven-thousand-dollar-a-month payments, and they will do anything to work them. *Anything.*" During the brutally cold January of 1994, for example, some loggers tried to work when it was forty degrees below zero. The lucky ones couldn't get their saws and skidders and harvesters started. The unlucky ones risked ruining their machines by running them when the oil was the texture of tar.

One way to reduce the pressure on the loggers, Dick suggests, is to convert to an hourly pay system. Andy bristles at the idea. "I'd sooner work at Shop-and-Save bagging groceries," he says. "What's the difference? You'd take away all my incentive to work if it was hourly." It's not the piece-rate system itself that makes it so hard to make money, he says, but overzealous OSHA regulations, poor planning by landowners, rising costs, and the stagnant level of the rate paid by the contractors. Andy also resents Treeline's incentive program that offers a bonus to anyone who cuts over 100 cords in a week. While Andy has cut as many as 122 cords in a week, he averages around 40.

"It's like they're saying, 'If you were a real woodcutter you would cut one hundred cords a week.' Well, I worked in a place where there were families of three generations of woodcutters, and there wasn't one of them who had cut over fifty cords a week in his life. They were well-loved employees. But here it's never enough. We're always looking for quantity."

Dick Slike doesn't make excuses for the incentives; Treeline must produce wood or lose its contracts. But he agrees with Andy about the other pressures. He says it's a mistake for paper company foresters to leave so many on-the-ground decisions to the loggers. Ultimately, he says, the land will suffer.

"Andy has to worry that the landowner will be happy with what he's done," Dick says, "but to make any money he has to get all the wood out to the road as quickly as he can. Then OSHA comes in and says he has to knock down all the dead trees in the work site and cut perfect wedges into every tree so it will fall exactly right. Then the landowner doesn't want to build more roads. So once again, it's all thrown on to the guy whose money has to be made in production." Dick says that foresters for the paper companies should measure off each strip in the strip cuts, try to reduce skidding distances by planning roads more carefully, and mark each tree they want cut in the basal area harvests. "To me, that's their job. That should not be Andy's decision."

Dick made this argument at Treeline's annual meeting with International Paper. The foresters listened and nodded. "But at the end," says Dick, "their answer was, 'This is what we have for you to cut. Take it or leave it.' Now these are good fellows and they're good foresters. I think they have their contractors' interests at heart to the extent that their corporate higher-ups will let them. But someone at corporate headquarters crunches the numbers, and that's it." Among the effects of this cost-cutting, Dick says, has been a reduction in the number of foresters in the woods. He says the same is true of other paper companies that Treeline contracts with in the Lincoln area, and that the situation is even worse with land speculators, who rarely use foresters at all. They just contract with Treeline to cut every tree they can sell.

Andy sighs. He and his wife have just finished building a house in Lincoln, and they have two young children. Someday Andy would like to buy a woodlot and manage it himself. But for now he has no desire to roll the dice against the crushing insurance costs and marketing risks of becoming an independent contractor. With Treeline handling the administrative headaches, he's happy to be a "logger logger." Even so, he questions how long the land can sustain the impact of what he is sometimes told to do.

"We make desperation attacks on these stands of wood like we are never coming back," he says. "We've gone through some great little rock maple stands and just 'whooooomp.' Gone. Devastated them. They would have been great stands of wood to thin. But instead they find a little softwood regeneration underneath, and tell us to whack off all the hardwoods. I just don't get it. It's like a plague. Is that going for the long haul? I don't know. I just take what I can get."

9

The Industrial Forest

You needn't look far to find critics of forestry in Maine. Two decades of rising harvest levels have left imprints on the land that, in terms of a human lifetime, seem permanent.

"You can fly over Maine and never leave a clearcut for more than a half hour," testified environmental activist and philanthropist Charles Fitzgerald of Atkinson at a 1994 Northern Forest hearing. "They've ruined the forests in the state of Maine," logger Dana Marble of Topsfield told us. "There are places the paper companies have cut where I don't think we'll see another harvest for sixty or seventy years." Mitch Lansky's 1992 polemic *Beyond the Beauty Strip* declared commercial forestry in Maine to be an economic, scientific, moral, and cultural failure. Lansky implicated not only the timber industry, but government, academia, and environmental groups. At hearings held by the Northern Forest Lands Council (NFLC) in spring 1994, many Mainers, including several foresters and loggers, lamented the state of the woods. Some compared commercial forestry with commercial fishing off the New England coast, which in the early 1990s essentially collapsed from overfishing and inadequate conservation policies. Others described how childhood memories of deep mossy woods have given way to grim images of gravel highways and clearcuts.

"There is a concern," says natural resource consultant and NFLC member Jerry Bley, "that we are running out of trees."

Maine's is a hard-working forest. Its 17.5 million acres of woodlands, 90 percent of the state's land base, make it the most heavily forested state in the nation. Nearly half of this forestland, 8.1 million acres, is owned by eight Fortune 500 paper corporations and by sawmills or other manufacturers. That's the nation's highest percentage of "industrially" owned forest, so-called because the companies also own manufacturing facilities. Large "non-industrial" landowners that make their money selling wood to industry, such as the Baskahegan Company, own another 3.1 million forest acres in tracts larger than 5,000 acres. Approximately one hundred thousand small landowners control 5.4 million acres, and the rest, 900,000 acres, is public – at 5 percent, the smallest percentage of public land in the Northeast. In addition to managed timberland, paper companies and other commercial forest owners also own several million acres of ecologically important wetlands and bogs. Ten million acres of commercial forestland are in the unorganized townships, where there is no local government and few residents. In all, as much as 98 percent of Maine's forest is available for timber management.

Wood harvested from these lands goes to thirteen pulp mills and eighteen paper mills, fifty-eight large sawmills and hundreds of smaller ones, dozens of manufactured-lumber plants, and twenty wood-fired power facilities. Secondary wood manufacturers – including producers of dowels, toys, toothpicks, canoe paddles, furniture, and pallets – run in the hundreds. Landowners and wood brokers export high volumes of wood to Canada and ship significant amounts overseas. The Maine forest industry has increased its production capacity steadily over the past two decades; today it stands capable of producing 12,000 tons of paper and 2 million board feet of lumber every day. Paper company dams control long stretches of the state's 31,000 miles of rivers. In many parts of the state, private logging roads are the only way of getting from place to place.

Poor data make it impossible to pinpoint the exact volume of wood cut and acres of land harvested before 1990, but the trend rose steadily from 1970 through the late 1980s. In 1991, after state law mandated rigorous landowner reporting, the Maine Forest Service estimates that landowners harvested wood from 437,292 acres. Excluding "inoperable" lands such as very steep terrain and certain kinds of bogs and wetlands, that means 2.5 percent of the forest was harvested in some manner. At that harvest rate, a young woman graduating from high school this year could see virtually the entire state, every square mile, logged to some degree before she reaches retirement age. Meanwhile, the land that was logged while she was growing up will bear the visual effects at least until her children are in high school.

Confronted with these figures, some may ask, "So what?" Logging has been part of Maine's culture and economy since long before it was a state. Forest products comprise the state's most important manufacturing industry, accounting for some 40 percent of all manufacturing shipments and employing twenty-five thousand people, a quarter of the total manufacturing work force. Paper mill jobs are the highest-paying blue-collar positions in Maine, and lumber mill jobs rank fourth. Another twenty-five thousand Mainers work in forest-based tourism, relying on privately owned timberlands as *de facto* recreation areas. In all, forest-based businesses employ 12 percent of Maine's workers and account for 11 percent of total state payroll. Most residents will rarely hear a chainsaw in the places in which they camp and hike and fish – which, again, are usually on private, working timberland. For decades Mainers have generally accepted that theirs was a working forest, that its extraordinary resilience could always rebound from heavy use.

Yet many people are now growing uneasy. They see signals that long-term economic, aesthetic, and biological integrity is being sacrificed for short-term profits. Meanwhile the traditional buffer against criticism, the industry's enormous economic impact, is breaking down: While harvest and production rates climbed steadily from 1970 to 1990, forest-based employment, especially in the woods, fell. The most visible targets for criticism have been large clearcuts, where landowners cut down all the trees in a given area. Clearcutting increased dramatically as a method of forest management during the 1980s.

"With employment shifting down and uncertainty about what's happening with forest practices, I think people are changing," says the Maine Audubon Society's Mike Cline, a forest ecologist and a former researcher for International Paper. "The signs are more apparent. It's that legacy of two decades of clearcuts that are there on the landscape and don't grow back for fifteen or twenty years. Wilderness-wise, it's going to take a hundred years.

"It must have been boring to be an industrial forester in the last decade or two," Cline muses. "You essentially go out and mark the perimeter and cut everything inside it. Then you go back and see if the regeneration survived."

The word *forestry* first appeared in dictionaries in 1860, defined as the "art of forming or cultivating forests." Over the next several decades, this "art" merged with a growing understanding of the biological processes of natural forest succession. Trees grow, drop their seeds, die, and fall. In their place, nurtured by varying amounts of sun and water, and fed by nutrients from soil, rotting leaves, fallen branches, and downed trees, the next generation establishes itself. Species follow one another in largely predictable patterns based on differing tolerances to sunlight and other conditions. Disturbances such as pestilence, disease, fire, drought, and wind accelerate and change the succession process. Such disturbances can affect single trees or entire forests.

The essential premise of scientific forestry was that this natural succession could be controlled and sustained to suit human needs. This view of trees as an agricultural crop was one of two reactions to the heavy timbering that occurred at the end of the nineteenth century. The other reaction was a clamor of support for public parks. "The preservers wanted to save the trees, the conservers to use them more wisely," writes historian Stephen C. Fox. As personified by the strained friendship between John Muir, who helped inspire and lead the early preservationist movement, and Gifford Pinchot, the "father" of scientific forestry and the nation's first chief forester, the philosophies of preservation and conservation grew apart around the turn of the twentieth century, each taking its own proponents. President Theodore Roosevelt sided with Pinchot, who advocated forest management as a way of providing "the greatest good for the greatest number for the longest time." Forestry thus emerged at the turn of the century with some impressive credentials: It was utilitarian; it was egalitarian; it was scientific; and it enjoyed political support.

By 1910 all four states in the Northern Forest had created forestry departments, and Pinchot's U.S. Forest Service had been elevated to a place of distinction in the Department of Agriculture. (This bureaucratic niche symbolized the government's view of wood as a crop, and institutionalized the treatment of National Forests as timber reserves rather than as parks; the National Park Service, by way of contrast, is housed in the Department of the Interior.) Many timber companies became avid supporters of forestry. This was partly because the new state forestry departments were also controlling forest fires, but, more fundamentally, because progressive lumbermen realized that forestry could sustain profits over a long period of time; they realized that the science could move the industry out of the "cut-and-move-on" phase of the nineteenth century. By the 1910s, many timber firms were researching tree growth and planting carefully cultivated seedlings, further underscoring the parallels between forest management and agriculture.

Those early days of forestry saw the development of what has remained the utilitarian standard for "good" forestry: the concept of sustained yield. This is the ability of a stand or an entire ownership to produce the desired volume of the intended species of wood for an indefinite length of time. In the environmentally conscious 1960s and 1970s, "multiple use" was added to this equation. Timber was still the goal, but foresters were urged to consider the impacts of their management on other forest "uses" such as wildlife (especially game species), scenery, recreation, and water supplies. Minimum guidelines to protect surface water, known as "best management practices," were written into state statutes and into professional standards for foresters.

During the 1970s and 1980s, forestry attracted thousands of young people who enjoyed being outdoors and physically active, and who wanted to bring the intellect to bear on complex problems of biology and economics. Particularly intriguing was the notion of using the best knowledge available to mimic, in a productive way, the processes of nature. What wasn't immediately apparent to these future foresters was the extent to which economics might conflict with – and at times override – pure science.

One such forester was John Steward, a Maine native who took his first job in 1990 as an operations forester with Champion International Corporation. He was assigned to help manage the company's land in northern New Hampshire and Vermont, planning timber harvests and supervising loggers, many of whom had been cutting wood for longer than he had been alive. After some rough initiations, including one case where he had to shut down a crew that had used a backhoe to alter a stream bed, Stewart began to enjoy working with the loggers. Yet he wasn't applying much of what he had learned at the University of Maine.

"The part that got a little wearing was having to meet a bottom line," Steward says. "That was kind of a big shock. 'You mean I can't do a shelterwood here because there is too much valuable veneer? Hmm . . .' But you learn to adapt."

One of Stewart's colleagues, who left industrial forestry in frustration after five years, said that Champion's foresters referred to themselves as "mud cops."

"We had so many machines to watch, so much territory to cover," this forester says. "I'd go to a job, walk up and down the road, and basically make sure they weren't driving the skidders in the stream. That's about it – keep them out of the water. We were cutting like there was no tomorrow, going back into areas we had just been in three years earlier. It was never a question of going to the foresters and asking us how much we could get out of the woods and still leave some for tomorrow. It was just like any other business where you go to the shelf and pull out what you need for the year without thinking about how to replace the inventory." A low moment came when one of the foresters suggested at a staff meeting that the company replace its foresters with accountants in skidders.

By 1993, Champion's heavy-handedness in northern New Hampshire and Vermont was raising eyebrows outside the company as well. The concern stemmed partly from journalist Richard Manning's 1992 book *Last Stand,* which reported that Champion was consciously liquidating all of its timber in Montana. This revelation was followed in 1993 by a *New York Times* report that the denuded Montana land was for sale. Champion suddenly appeared to be the archetypal cut-and-run operation, and even some in the industry started questioning the company's commitment to northern New England. In fall 1993 Champion publicly acknowledged that its cutting rate in New Hampshire and Vermont was not sustainable and pledged to "drastically reduce" the harvest. Shortly thereafter, Champion re-organized its northeast timberlands division so all decisions would be made from a central office in Maine, where the company owns some 730,000 acres. The company announced it was redoubling its commitment to "stewardship." A team of senior officials hit the road to meet with environmentalists, wildlife biologists, and sportsmen to talk about how Champion could better accommodate the public's demand for "non-timber values." While some shrugged this off as pure public relations, others realized that the company was adapting for the most basic reason of all: its pocketbook. "We feel," said a highly placed Champion forester, "that we are losing our prerogative to manage our land."

By then, John Steward had been reassigned to Champion's regional headquarters in Bucksport, Maine. Now he splits his time between supervising the company's recreational camp leases and assisting with planting trees and spraying herbicides to control hardwood growth.

"I agree with a lot of the points that some of the environmentalists bring out," he told us one day. "But they're talking about textbook silviculture." He was sitting outside a Champion district office, waiting for the breeze to die down so he could call the helicopter back in to resume spraying herbicide on a plantation of black spruce. "I'd love to do textbook silviculture day in and day out. But when you get into the economic end of it you have to find that compromise. A lot of times the critics are not willing to look at the economic end of it. They say, 'This is the way it should be. How come they're not doing it?'"

Many complex factors were behind the heavy, revenue-driven logging that John Steward and his peers found themselves trying to supervise in the 1980s and early 1990s. These factors included the creation of separate timberland divisions as stand-alone businesses within some paper companies; the economic maturity of the region's forests; Wall Street's obsession with short-term cash flow and double-digit returns; the fear of hostile takeovers; mechanized logging; rising land values; and escalating expenses, including the cost of paying property taxes and following environmental and safety laws. Pulp prices roughly doubled during the 1980s, resulting in a flurry of investments in mills to take advantage of the market. Paying for the improvements meant running the mills hard. The same was true with mechanized logging equipment.

Exacerbating and contributing to these factors was the continuing reaction to the devastating outbreak of spruce budworm in the mid-1970s, even though the outbreak ended in 1985.

"Forestry has always been impacted in Maine by cycles of the budworm," says Pat Flood, chief of forest management for

International Paper in the northeast. "As those cycles come and go, the forest characteristics change. And then there is a tremendous rush to cut wood because it's either dying or it's dead. It's the very ecosystem that we deal with in Maine. We call it the spruce-fir forest, but you could call it the spruce budworm forest, too, because that's really how the forest perpetuates itself."

"It was forestry under siege," says an official for Champion. "From 1976 to 1985, the budworm taught us a lesson. We were in a panic." As can happen in a panic, the lesson that emerged – to increase the harvest rate and cut trees before they were threatened by the pest – posed its own problems.

"The budworm led to a love affair with the clearcut," says consultant Jerry Bley. "Long after it was gone, we were hearing foresters say, 'The budworm made me do it.' We're *still* hearing that."

Some defended clearcutting, saying the practice imitated the impacts of fire and hurricanes and pestilence, and was necessary to achieve silvicultural objectives in certain conditions. While both explanations were true to some degree, the rationalization fell short given the number of acres involved. Although harvest data gathered by the Maine Forest Service before 1990 are considered to be unreliable, the agency does stand by the earlier figures for purposes of estimates and tracking trends. According to Forest Service reports, landowners clearcut approximately 100,000 acres every year through the 1980s, with highs of 138,000 acres in 1987 and 145,357 acres in 1989 (nearly half the acreage harvested by all methods that year). Fires and windstorms, critics pointed out, just don't wipe out a hundred thousand acres in geometric patterns every year. Large-scale clearcutting was not an emulation of nature, but a cheaper way to cut, the best way to salvage budworm-infested stands, and the fastest way to convert a hardwood or mixed wood site to the softwood species desired by many large landowners. Something less than ten percent of the clearcuts were replanted with softwood seedlings, usually spruce. A larger portion were treated with herbicides or hand-operated brush saws to control hardwood growth from competing with the softwood. The rest were left to regenerate naturally. Unless naturally occurring softwood seedlings were well established before the clearcut, these sites often came back to dense tangles of raspberries, poplars, gray birch, and pin cherry trees. In addition, areas that were clearcut in strips or patches often created wind tunnels that blew the surrounding forest down, creating larger clearings than the landowners and foresters had intended.

By the late 1980s, objections to large clearcuts were growing, both inside and outside the industry. Some foresters and commercial woodland managers feared that political backlash against the large clearings would erode public support for all forms of forest management. Ecologists warned that clearing mixed forests and converting them to stands of pure softwood could accelerate the next budworm outbreak. (Some biologists believe that although budworm outbreaks have occurred periodically for millennia, the last three – in the mid-nineteenth century, the 1910s, and the 1970s – have been exceptionally intense.) Environmentalists were becoming more vocal, and aerial photographs of sprawling clearcuts in Maine started appearing in regional publications.

"There's so much biological legacy lost when you clearcut something," Mike Cline says. "If you hack the forest down when it gets to economic maturity, you lose the complexity – compositionally, structurally, and functionally. It just doesn't function the same." Cline and others argued that in the highly recuperative forests of Maine, treating trees like crops to be cleared and then starting over with raw land simply didn't make sense. As Cline has put it, "The best way to grow wood is on existing trees."

In 1986, Mike Cline, Pat Flood of International Paper, Roger Milliken, Jr., of the Baskahegan Company, Ted Johnston of the Maine Forest Products Council, and several other conservation and forestry leaders in Maine created a "Forest Forum" of industry, environmental, government, and academic interests to discuss clearcutting and other issues. In 1989, the group submitted to the state legislature the Maine Forest Practices Act to restrict clear-cutting, strengthen reporting requirements for logging activities, and increase staffing at the Maine Forest Service. The law defined a clearcut as a harvest that left less than 30 square feet of basal area per acre, which is not a lot of trees. (Basal area is the horizontal surface of a tree at 4½ feet off the ground. To visualize what 30 square feet of basal area per acre looks like, imagine looking straight down at a square acre of land, which measures 44,100 square feet. If all but 30 square feet of basal area had been cut, your acre would be very thinly sprinkled with about thirty trees of 14 inches in diameter, or seventy five trees of 6 inches in diameter.) For any harvest over 35 acres that met this definition, the landowner had to leave 1.5 acres uncut for every 1 acre cut, provide a "buffer" of uncut or lightly cut forest between clearings, report the location of the harvest to the Maine Forest Service, and write a management plan. (Cline, a key architect of the legislation, smiles at this last stipulation. "What's the management plan for a clearcut? You clearcut it and see if there are any seedlings left. No? Plant it . . . Yes? Leave it alone.") Successively wider buffers were required between larger clearcuts up to an absolute limit of 250 acres.

In the wake of the new law, the industry's reliance on large clearcuts in Maine dropped significantly. According to landowner reports compiled by the Maine Forest Service (reports which the agency now considers to be more accurate but still not precise), 59,602 acres were clearcut in 1992, or 12 percent of the land that was harvested in some manner.

"We believe we can accomplish our objectives and still stay under the thirty-five-acre threshold more often than not," says Pat Flood, whose company clearcuts approximately 5,800 acres in Maine a year, or one-fifth of the total area they harvest. "That's the spirit of the law, and we're trying to do that." Flood says the company still clears some tracts larger than 35 acres, but never exceeds the 250-acre maximum.

Nonetheless, Flood says, following the regulation puts additional pressure on already heavily worked foresters. "There are more pressures today on a forester's time," he says. "There's a lot of pressure to practice the silviculture that they think is right, create appropriate levels of earnings for our corporation, and keep their harvests at an appropriate level. That's a hard thing to do. A *hard* thing to do."

The Maine Forest Practices Act reduced the number of large clearcuts, but has done little to convince critics that commercial landowners can always manage the difficult balancing act that Flood describes. "People were angry about clearcuts before we had the clearcutting law, but nothing to the extent they are now," Cline says. "People are just ripped. It seems so crazy, because clearcuts have gone down. But that cumulative effect has caught up."

Since the Maine Forest Practices Act essentially reduced the number of trees that could be cut per acre, it appears that more acres are now being cut. In particular, some people fear that the act has led to more "high grading," or cutting all the best trees, a practice that allows landowners to avoid the definition of a clearcut. The problem is, a careless partial harvest, especially high grading, can cause more damage than a well-planned clearcut. Says Maine forest economist Lloyd Irland, "The surge in clearcutting was highly visible and made a lot of people mad. But we could ban clearcutting altogether and still ruin the forest." This leads back to Jerry Bley's question: Clearcutting or not, is Maine running out of trees?

When considering this most basic question, forest researchers use the term "timber budget." The timber budget for any given area is calculated by determining the net annual growth (total growth minus natural mortality) and subtracting the annual harvest. Back in 1920, after a century of heavy cutting and a decade of pestilence, Maine's timber budget was probably deeply in the red – one estimate suggests that twice as much wood was dying and being cut as was growing. The nation's lumber industry had largely moved west, and economists were predicting that the Northeast's pulp and paper industry would fail within thirty years. Then in the following decades, many owners, such as the Baskahegan Company, let their land sit and recover. By 1970, Maine's timber budget was back in the black with a mature, mostly evenly aged forest that had regenerated from the earlier exploitation. Then the new budworm outbreak and the wave of investment started the heavy cutting all over again.

The exact status of today's timber budget is impossible to determine, because the U.S. Forest Service has not done a comprehensive inventory of the region's timberlands since 1982. Nonetheless, the 1992 U.S. Forest Service report, "Forest Resources of the United States," in which actual inventory figures from 1982 were projected forward ten years, is somewhat illuminating. The report estimates that there were 24.1 billion cubic feet of standing wood in Maine's forests in 1991, with almost twice as much softwood as hardwood. After natural mortality, the report projects, the forest's net growth was .51 billion cubic feet, or about 2.1 percent. That year, landowners harvested .45 billion cubic feet, 2 percent of the original volume, or – more significantly – 88 percent of the year's net growth. (By contrast, using figures from the same U.S. Forest Service report, landowners in the three other Northern Forest states harvested about 62 percent of net growth.) In oversimplified terms, think of Maine's forest as one tree 100 meters tall. Between 1991 and 1992, after some natural dieback, the tree grew to 102.1 meters. Paul Bunyan then came by and sliced off the top 1.8 meters, leaving 0.3 meter more than there was at the beginning of the year. That one-third of a meter – on a 100-meter tree – is the actual growth, or the timber budget for 1991.

In theory, as long as this happens – as long as more wood grows than is lost to mortality and logging every year – a landowner should be able to harvest wood indefinitely. But there are problems with this theory. First of all, estimating the timber budget, at least as of 1995, is all guesswork, for the U.S. Forest Service has been tardy in completing its comprehensive "decennial" forest inventory of the northeast (Maine's new figures are due in 1996). This lack of vital information forces economists, foresters, and biologists to rely on projections rather than hard numbers. Some observers, including the woodlands manager of a major paper company, worry that the next Forest Service study may show that the harvest rate of some species is outpacing growth, not just in Maine but in other parts of the region as well. "We talk about land-use planning, yet we don't know how many goddam trees are out there," this official says. "We don't know how fast we're cutting, and we don't know how much we're cutting."

Second, at present much of Maine's actual growth is concentrated in young, fast-growing hardwoods and regenerating clearcuts, while the harvest is concentrated largely in higher quality, older stands of softwood. To some experts, this means that the forest is too young, and that quality is declining. Lloyd Irland endorses U.S. Forest Service estimates that 48 percent of Maine's woods are over-stocked with small, even-aged trees that need to be thinned. Until these stands are thinned, average log size and the percentage of mature forests, especially spruce, will continue to decline for many years. Silviculturist Robert Seymour of the University of Maine predicts shortages of high-quality mature red spruce beginning early in the next century and lasting for ten to twenty years. Further, if estimates are accurate and landowners are cutting 88 percent of the actual growth every year, they are leaving an uncomfortably small timber budget to absorb a major infestation, storm, or other natural disaster. Consequently, Irland, Seymour, and others have long urged landowners to thin more young stands rather than just harvesting mature stands. That way, they can diversify the age balance and spread out harvest pressure and growth more evenly.

Finally, and most important in the view of many experts, the timber budget is only one means of gauging forest health and a faulty one at that. Simply counting trees does not consider their larger role as integral components in a complex ecosystem that we know remarkably little about.

Reed Noss, a leader in the emerging field of restoration forestry, writes, "If our goal is only to maintain an approximately even flow of wood products, then we have a seemingly easier task than if we have to worry about sustaining the food webs and nutrient cycles that maintain soil productivity. Of course," he points out, "in the long run we must think about maintaining soils and ecological processes if we want a sustained yield of wood products."

As Mike Cline puts it, "If the companies are looking solely at timber, at farming fiber, they'll probably get through. It's certainly not eloquent forestry, though."

Eloquence may seem a tall order for foresters. Yet Noss's and Cline's comments underscore that, after decades of accepting sustained yield and multiple use, society is demanding more from forestry. Wildness, biological diversity, backcountry recreation, and aesthetics are no longer considered byproducts of forest management, but essential goals as important as fiber and lumber production. This adjustment is central to the emerging approach known as "new forestry." Underlying new forestry's tenets is the much broader concept of "ecosystem management," which considers the ripple effect on surrounding lands of manipulating any given stand. Rather than managing for sustained timber yields, the argument goes, foresters should try to preserve and enhance the intricate biological processes and relationships that contribute to overall ecosystem health. Do that right and the forests can produce a range of ecological, economic, and social benefits, including timber. In practical terms this means diversifying species and age in the forest; extending harvest rotations; leaving organic material in the woods to rot; letting some trees and even entire stands grow old and die; using prescribed burns to simulate the unique effects of fire; reducing the use of backcountry roads; expanding wildlife habitat; and leaving large tracts of forest intact. What's critical is not what you cut, new forestry advocates say, but what you leave. Especially bigger trees.

There are economic as well as ecological reasons to lengthen harvest rotations and cultivate older, bigger trees. Prices for lumber are soaring as a result of global demand, overcutting in other parts of the world, and the tightening of softwood supplies in the Pacific Northwest. Industry observers see tremendous opportunity in these trends for Maine's large landowners. Cline and Irland, for example, argue that the paper companies could meet a greater portion of their pulp requirements by buying it from elsewhere, and by using more recycled materials. Then the companies could devote less land to growing pulpwood on short rotations and more land to growing large-diameter sawlogs on longer rotations.

The challenge is to prevent cyclical waves of demand and new technologies for harvesting and processing wood from periodically impoverishing one or more components of the forest. This will be increasingly difficult as the global timber demand continues to expand. "The world," claims a leading log exporter from Boston, "is coming for New England's forest."

One prominent idea for preventing such impoverishment is the "triad," a technique developed by Robert Seymour and Malcolm Hunter of the University of Maine. The triad concept would concentrate high-yield forestry in certain areas where soils and other conditions are appropriate; identify and set aside biologically unique areas as ecological preserves; and promote "new forestry" on the remainder of the land. The preserves would protect habitat for plants and animals that are sensitive to disturbance, and would meet the needs for backcountry recreation. More significant, they would provide an ecological benchmark against which forest health in the working forest could be measured, and create a genetic bank to hedge against impoverishment in the intensively managed areas.

Many conservationists prefer a "diad," which leaves out the intensive management component. Mike Cline and Jerry Bley admire the forestry of the Baskahegan Company and Seven Islands

Land Company (managers of one million acres of land owned by the heirs of John Pingree), both of which prefer careful shelterwood and selective harvests over clearcuts, leave biomass in the woods to feed the next generation of trees, aggressively market low-quality wood to help pay for thinnings, and employ sophisticated, lower-impact logging equipment. But such an approach requires that landowners reduce short-term profit expectations in favor of longer-term gain. Privately held companies such as Baskahegan and Seven Islands can more easily afford to exercise this restraint. It's a different situation for publicly traded corporations trying to feed their mills, provide consumers with low-cost goods, and pay quarterly dividends to thousands of shareholders.

Some feel this is the most serious problem of all.

"At the heart of a lot of this," says Jerry Bley, "is the basic incompatibility of our economic system with forestry."

In 1994, International Paper started running a series of artful double-page advertisements in national magazines, including the *New Yorker*. Image ads for paper companies are nothing new; Weyerhauser, Georgia-Pacific, and many other firms have long used them to position the companies as corporate Johnny Appleseeds, planting trees in neat rows across the land. The new IP series was a little different, for it directly addressed the conflicts and cross-purposes that underlie Bley's comment.

Illustrated with a large sepia photograph of a blanket-shrouded child wandering in the woods, the ad read, in part, "Though we're one of the largest private landowners in America, we know it's not only our land . . . We know there are millions of people who rely on our forestland for different yet important reasons. Some need lumber to build homes. Some need tons of paper to run their businesses. Some just want a place to go camping or fishing. These are not complicated desires. But satisfying them involves complex personal, business and global issues." The ad encouraged readers to call Dr. Sharon Haines, toll-free, "to talk further." Haines, a renowned soil scientist and International Paper's manager of natural resources, has become a timber industry leader in understanding biological diversity and ecosystem management. She was active in the Northern Forest debate, and chaired the Society of American Foresters' Task Force on Sustaining Long-Term Forest Health and Productivity – a candid discussion of ecosystem management and new forestry that generated much controversy within the profession. She is the grandmother of the child in the photograph and the company's top spokesperson on land management.

Pat Flood, for one, welcomes the attitude adjustment that Haines personifies. "Fundamentally, there has been a movement in our company to do kinder, gentler forestry," he says. "We're learning how to do things differently. We're trying to blend aesthetics and landscape-scale management together so that in planning our harvests we're doing them not just to accomplish silvicultural objectives, but so some of these other objectives are met. We're very conscious of public opinion."

Pat Flood has worked for IP since he graduated from Syracuse University in 1974. He spent the first thirteen years in upstate New York as an associate forester, project forester, regional timberland manager, and finally manager of the company's extensive holdings in the Adirondacks. In 1987 he was promoted again to manager of forest management for all of the company's land in Pennsylvania, New York, and northern New England. Some 970,000 acres of this land are in Maine, as is Flood's office.

To move toward a gentler forestry, Flood says, IP foresters in Maine and elsewhere are leaning away from big clearcuts and planning more shelterwood cuts, pre-commercial thinnings, and basal area selection harvests. The foresters are also taking courses in landscape management to minimize the visual impact of harvest areas by, among other methods, planting grasses in disturbed roadsides and instructing loggers to push shorn limbs and branches back into the woods. Flood acknowledges that this approach sometimes requires the loggers to do more work for which they are not paid. "We have a few less foresters, we have a lot more responsibility, and a lot more area to cover," Flood says. "I don't think we need to make apologies for that, but we should probably commiserate a little more with the loggers. Because we find the same things happening to us."

This new direction not withstanding, IP still plans to cut a lot of wood in Maine. In 1993, the company cut 466,000 cords from about 30,000 acres in the state. Flood and his boss John Cureton estimate that the annual harvest will drop a bit between 1994 and the year 2000, to an average of around 436,000 cords. (That's down from a high of 650,000 cords in 1987.) Maine's forest grows at an average rate of 0.4 to 0.5 cords per acre per year. International Paper owns 970,000 acres in Maine. If the growth rate remains constant and the projected harvest doesn't change, the company will be cutting close to 100 percent of the overall growth every year.

Responding to the concerns of some experts that such cutting levels will eventually produce a gap when the existing mature timber is gone and younger stands have not yet economically matured, Flood says, "We've had Bob Seymour here giving us his lecture, and we understand what he's saying. We're trying to even out those peaks and valleys with precommercial thinning, planting, and herbicides so we don't have this long wait for the young timber. All those things will help fill those gaps so we can bring the forest back to a balance that we're comfortable with. We're trying to make our lands more productive." Flood says that as the lands become more productive, capable of growing more wood per acre, the harvest level may rise somewhat early in the twenty-first century.

"Folks think that the harvest level is set by some boogeyman in East Jesus, Somewhere, who says 'You guys will cut this much wood,'" Flood says. "That's not the way it works. We start from the bottom up and determine what the forest is able to sustain. We don't call it the 'sustainable' harvest. We call it the *appropriate* harvest – the right level for the forest given the forest's condition.

That's a number that's generated here, based on biology and nothing else. We reserve the right to adjust that number, but only within a certain framework."

Land ownership, Flood says, has a lot of challenges. He says large corporations like International Paper face the same problems that all landowners face, only the problems are far bigger. He also jokes that, as a forester, he is "deadly jealous" of Chuck Gadzik of the Baskahegan Company, who works under less-demanding pressures to produce high short-term revenues.

"It's hard for anybody to take into account all the considerations when you're harvesting 30,000 acres a year, because you have 30,000 potential screw-ups," Flood says. "We don't practice perfect forestry, but our people are really working hard. We're trying to learn. But at the same time we all have responsibilities to keep conducting business."

10

Sleeping with the Elephants

Forest City is 15 miles from Baskahegan Lake, on the Canadian border in northeastern Washington County. The town sits at the tip of a peninsula that divides two long lakes, Grand and Spednik, which separate Maine from New Brunswick. Half a mile from the customs station on the American shoreline of Grand Lake is Wheaton's Lodge, a cluster of ten tan cabins and a small lodge. Red squirrels twitter in the tall, pruned pines that shelter the camp. Cars parked beside the cabins are from New Jersey, Massachusetts, New York, Pennsylvania.

On a chilled, misty morning in late June, a few guests are making their way from their cabins to the main lodge for breakfast. Inside the pine-paneled dining room, the guests find their tables and pick up simple typed menus. A waitress takes orders for eggs and bacon and pancakes and oatmeal. While the guests eat, six pickup trucks pull one by one into the driveway. Each truck pulls a trailer with a green, square-sterned canoe.

At 7 A.M., the screen door of the main building bangs as Dale Wheaton goes out to meet his fishing guides. Dale is forty-three, six feet tall and broad-shouldered, with red-blond hair and mustache. In each sunburned hand he carries a wicker picnic hamper with a guide's name written on masking tape on the handle. As the guests finish their breakfasts and return to the cabins for fishing rods and windbreakers and caps, Dale gives each guide a few details about the client, or "sport," that Dale has assigned him. They nod and peek into the baskets to make sure the cook has packed the right lunches.

137

It takes another half hour for the guests to meet their guides and get themselves and their gear into the right trucks. Dale stays in the dooryard, joking with everyone and directing traffic until the last truck drives away. Then he backs up his own Chevrolet to a shed, attaches a trailer with a Grand Laker canoe identical to those hauled by the guides, and fills the back of the truck with life vests, tackle boxes, fly rods, gas tanks, and a 25-horsepower Evinrude outboard. We climb into the cab. Dale sticks a nine-inch cigar in his mouth. "Ready?" He turns the key and pulls out of the driveway.

A few miles past the dozen multicolored houses that are Forest City, Dale turns left onto a gravel logging road. On either side of the road, the land is thick with waist-high gray birch and poplar, the remains of a Georgia-Pacific clearcut. "They did a nice job in here," Dale says over the rattle and thump of the rugged road. He's so deadpan it's hard to tell at first how far his tongue is in his cheek. "I'm glad they didn't leave any trees. I hate it when they leave a few. It's like being in Westchester County and seeing some guy who won't mow his lawn." He steers around the potholes.

We pass out of the clearcut and into the still intact forest along Spednik Lake until the road ends, essentially, in the water. Dale floats the canoe, parks the truck, and gets us settled in the boat. Then he steps in and tugs the outboard to life. As he motors across the cove, his untucked plaid shirt billows out from under a blue down vest. He wears a faded corduroy baseball cap and is still clenching the unlit cigar in his teeth. The wide canoe planes easily, sending a gentle wake rolling through the water. We emerge from the cove and into the broads of the lake, where sky and water are the same color of flat pewter as far as we can see, broken only by a thin green ribbon of shoreline. There's not a house nor a boat in sight.

As we approach a cluster of boulders breaking the surface of the water, Dale throttles the engine down and shuts it off. The stern rises gently as our wake catches up. Dale picks up a paddle and points to a boulder. "There," he says. "Drop your popper right next to that rock." The first cast misses the boulder by ten feet. Dale grins around his cigar. "Try it again." The second cast is little better. The third lands within a yard of the target. "Good enough," he says. "Just let it sit. He's under there. You'll drive him nuts if you just leave it there." Dale strikes a match on the gunnel of the canoe and lights his cigar. A swirl engulfs the little yellow lure. "Hit 'im!" Dale shouts, pulling the cigar out of his mouth. A minute later, the fish is at the side of the boat. Dale pulls the fish from the water by hooking a finger under its mouth. It's a 12-inch smallmouth bass, green and fat. Dale works the hook out of the jaw and tosses the fish into the lake. It's gone with a flick of the tail. Dale wipes his hands on his jeans and takes a few strokes with the paddle. He points to a boulder where he wants the next cast to drop.

In the next hour we catch a dozen bass. Dale eases the boat through the boulders with soft paddle strokes, keeping up a running commentary on our casting. At one point, feigning exasperation, he leans forward and takes the fly rod. With a single strong back cast and a perfect flip forward, he sends the little popper thirty feet out. It lands six inches from the rock he was aiming for. A bass rises instantly. Dale grins and hands back the rod, its tip bending double to the water.

We keep no fish. Guides working out of Wheaton's rarely bring back more than two, even on a good day. And the fishing is often as good as it is today.

"That doesn't mean we never clobber a fish," Dale says. "We're not saints and we're not angels. But if we don't have an ethic, who's going to?

"It wasn't always like that, though. I grew up in a period when we killed more fish, honest to God. There wasn't any end to it, or that's what we thought. But we woke up. Not because anybody told us, but because it dawned on us. It just stood to reason that you couldn't kill your seed and still have your young, you know?"

Dale Wheaton was born in 1950 in Grand Lake Stream, Maine. Two years later, his mother, Ruth, and his father, Woodie, founded Wheaton's Lodge. Dale started guiding at fifteen and has never missed a summer since. Income from guiding put him through the University of Maine and graduate school in economics in England. He married in 1973 and moved with Jana, his wife, to the Bangor–Old Town area, where she taught school and he monitored water quality for the regional planning commission. They had summers off and lived in a cabin on Grand Lake so he could fish.

In 1976, Dale's parents sold Wheaton's Lodge and retired. Three years later, Dale and Jana quit their jobs and bought the family business back from the new owners whose marriage, Dale says, "hadn't survived it."

Woodie Wheaton was one of Maine's best-known guides, and Dale and Jana were determined to retain the flavor of his sporting camp. This meant, first, finding the right size: large enough to be economically viable, small enough to keep from becoming a strain. After sixteen seasons, they've got it just about right. The Wheatons aren't getting rich, but they're getting by. When the camp is full, thirty guests sleep in ten cabins, simple pine structures with screened porches and comfortable beds and Canadian gliding chairs covered in plaid. Breakfast is at 6:30 A.M., lunch is on the lake, dinner is at 6:30 P.M. Ten or twelve menus rotate through the season, with three choices on any given night. The food is plentiful and simple and served with unusual condiments like marinated corn relish salad and pickled beets. No alcohol is served in the dining room, but a bucket of fresh ice is delivered to each cabin at 5. After dinner, Dale is generally in the little sitting room at the lodge, talking about fishing or the careers of his guests. Many he knows well; a number of clients have returned to Wheaton's for twenty summers or more. Some bring their own boats, but most fish with guides. Dale urges clients to hire a guide partly so the camp can make additional money. But his insistence also reflects a sense of tradition and place. He seems to want to *introduce* people to the lakes, as you would to a person of respect and importance.

"The guides that I have, they learned to do it my way, in the same manner I had to do it my father's way," he says. "When you do it too much differently, you're no longer a traditional sporting camp. If you run a motel with a bunch of bass boats lined up at the dock and your customers go screaming around hunting for fish, you've got a wholly different product. We're a traditional camp, mainly because I'm a traditionalist in a lot of ways."

It doesn't take long to feel like you are a part of this tradition, because everything at Wheaton's is assuringly consistent. The lunches always come in the square wicker baskets. All the guides own Grand Laker canoes, many of them handmade. They use roughly the same kinds of lures and flies and rods and outboard engines, with just enough variation to provoke the teasing and one-upmanship that their clients expect. Visit Wheaton's once, and the following spring you'll get a four-page newsletter called "The Backlash." Dale and Jana fill it with little bits of arcane news that are meaningful only if you've been there: "Andy has a new yellow Lab pup, Lily . . . Nimrod has a wholly new bathroom, and bedrooms finished in knotty pine . . . The average smallmouth bass reaches maturity at three years of age."

Recently, among the charts showing the previous season's best catches and photographs of guests fishing, "The Backlash" has also run stories about the land and waters surrounding Wheaton's Lodge. "Spirited Meeting on Lake Levels," read one headline in the 1993 issue. In the accompanying article, Dale reported that area guides and others were trying to convince Georgia-Pacific to be more careful with the way the company operates its dams. Georgia-Pacific owns the flowage rights on Spednik and Grand Lakes, which are part of the St. Croix River watershed. The company's storage dams control the flow of water to power its hydroelectric dams downstream in Woodland, which run G-P's paper and lumber mills. The newsletter story reported that several guides had testified at a hearing. They wanted to be sure that G-P officials understood that severe fluctuations in water levels threaten spawning bass and aquatic plants.

In the 1994 issue, next to a piece about Dale's mother Ruth and her watercolor paintings, an upbeat article announced that the Baskahegan Company would be selling a conservation easement on 14 miles of Spednik's shoreline to the Department of Inland Fisheries and Wildlife. Several small islands, including Woodie Wheaton Island, named by the Maine legislature to honor Dale's father, would also be protected. "Spednik exemplifies much of what we all value about the north woods," Dale wrote. "World class fisheries, abundant wildlife including bald eagles and loons, a strong sporting tradition, and a wilderness character. There is hope we can protect this masterpiece for future generations."

Wheaton's Lodge is surrounded by land and water controlled by Georgia-Pacific, Champion International, and the Baskahegan Company. The companies own the shoreline, the uplands, the rights to control water levels, the guides' favorite lunch spots on the lakes. Some of the sporting camps in the area lease land for their facilities from the large landowners, and all rely on them for access to the lakes. These relationships go back generations.

While Dale was growing up, the large landowners often cut their forests harder than the camp owners and guides liked, but at least the ownerships stayed intact. From the late 1970s through the early 1990s, however, Dale saw all that change.

"You probably heard the expression 'sleeping with the elephants,'" Dale said one afternoon. It was one of the rare days when more guides than guests are at Wheaton's, and Dale has stayed ashore. "That's pretty much what the little people have done around here – they've slept with the elephants. But the elephants didn't roll over very much in bed, so you knew what your position was.

"Nowadays it's a different deal. Corporate policies can change instantaneously. Some are better, like Baskahegan, but even they aren't saints. All of the big timber companies have different policies with respect to their lands and how much revenue they want to generate . . . how much they're driven by the bottom line."

Dale is generally tolerant about logging, because he knows that the large landowners provide most of the region's jobs. Every once in a while, however, he comes across a large clearcut that shocks him. Such as the time he was out on East Grand Lake with his eleven-year-old daughter, Kim. They landed on the shore and walked through the intact forest on the edge of the lake. Several hundred feet in, the forest was gone. "I mean there wasn't anything," Dale says. "It was absolutely devastated. There wasn't even any bushes in there. I guess I had gotten used to it, but it was different seeing it through the eyes of a child. Kim just couldn't believe that anybody could do that. She thought she was living in the woods up here."

Dale says he believes that some of the timber companies recognize the folly of such abuse, and he tends to be forgiving as long as the land stays in forest management of some kind. It's when the large landowners stop cutting wood and start cutting up land for development that his passion rises. Lakefront values rose astonishingly in Washington County during the 1980s, with building lots selling for as much as $9,000 per acre. Some companies couldn't resist the temptation to sell these parcels, which were marketed as "Wilderness Getaways" in the *New York Times* and the *Boston Globe.* Lease lots, on which the company keeps the land but leases out the right to build a hunting or fishing camp, were also increasing. Dale is a little less alarmed about leases than outright sales, but it's a subtle difference.

"Once the lot gets carved out, from my point of view, paddling along the shoreline, it doesn't matter," he says. "You're asking if I want to be killed slowly or quickly. Either way, I don't want it to happen." Dale has little problem with Baskahegan's policies in this regard; the Millikens have not sold or leased any land for decades. He is less sure of Georgia-Pacific.

"When I was growing up, G-P never ever sold any land. And then in the 1980s, their policies took a one-hundred-eighty-degree turn. Now, everything that is not strategically part of their forest base is up for grabs. If there is something you wanted, and the price was right, they'd sell it to you." Dale's peevishness about Georgia-Pacific is tempered somewhat by the company's tremendous economic contributions, its sporadic philanthropy, and his personal relationship with some local company officials. Bob Chandler, the company's regional forester and a friend of the Wheatons, has been known to angle strip cuts so they can't be seen from the water, and to leave attractive stands of trees around lease camps. In 1992, partially at the urging of Dale and his brother Lance, who also is a guide, the company sold 532 acres of shoreland near Forest City to the Maine Department of Inland Fisheries and Wildlife.

Dale knows too much about economics to lose sleep wondering why a multinational paper company can't see things his way. "Maine," he says, "is just a small part of the picture for these companies, and these are complex dollar-driven decisions. It's a mistake for some podunk like Dale Wheaton to try to analyze them." To the other major players in lakefront development, however – the land speculators – Dale gives no quarter.

When timber companies sell their highest value lands for development, they rarely get involved in the nitty gritty of subdividing and marketing. That's usually left up to land brokers and speculators. Dale gets particularly furious at one breed of speculators he calls the "bad boys," many of whom are fellow native northern Mainers.

"The bad boys buy major tracts of land, cut every single merchantable piece of wood, subdivide the lots, and sell them to folks in Massachusetts who see a forty-acre tract of land going for a thousand dollars an acre and say, 'That's a hell of a deal.' The bad boys have absolutely no reverence for what that piece of property looks like, the people around them, or the disposition of it ten years from now. They don't care how many brooks they cross with the skidder. They don't care how much sedimentation comes off the hill. They don't care about the quality of the roads they build and leave. They don't care what the hillside looks like from a distance, or pay any attention to the zoning around them. Those guys are the ones who have stirred such animosity."

Dale sits up straight and furrows his brow. He counts the "bad boys" off on his fingers. "Herb Haynes. Hank McPherson out of Bangor. Patten Corporation. These guys are in it for the maximum amount of money it's going to generate between the time they buy the land and the time they get rid of it. There's one hell of a lot of land that's moved that way in the state of Maine."

Despite his anger, Dale has begrudging respect for some of these businessmen, because he knows that they keep a lot of local people employed. "Take Herb Haynes," he says. "Herb is a very intelligent guy. Very articulate. Very ambitious. Good entrepreneur. And if you were a wood dealer with a contract with a paper company, Herb's your man, because he's not going to frig up. You need your two thousand cords of spruce, Herb's got them sitting in the dooryard. Now I'm a businessman and I recognize you have to exploit your property in some way. But it's all in the time frame. The bad boys just don't care about the next generation."

A report compiled for the Land Use and Regulation Commission, the planning and zoning body for Maine's 10 million acres of unorganized territory, concludes: "The fact that new residential development is gravitating to the most valuable lake resources . . . has an undue adverse impact on the traditional Maine Woods experience. The numbers of relatively secluded, undeveloped lakes and rivers is diminishing. The question might be raised, how long before what we now call 'semi-wilderness' becomes 'suburbanized'?"

So far, the greatest percentage of shoreline development in central and northern Maine has been around the more popular and accessible lakes, such as Moosehead and Rangeley. In Washington County, a lot of land was subdivided and sold in the 1980s – nearly

18,000 acres, as much as 90 percent of it on the lakes – but not much was developed before the real estate market declined in 1990. The speculators are just holding on to these parcels, waiting for the market to improve.

Dale is sure that will happen. So when his out-of-state clients start talking about buying a parcel of shorefront on Spednik Lake and building a second home, he scoffs at the idea, or changes the subject. Dale discourages such talk for a reason that is both selfish and pragmatic: He'd rather have visitors stay at Wheaton's Lodge. He wants the business, and he believes that traditional sporting camps create less permanent impact than individual vacation homes lined up along the lakes.

"I got the crap scared out of me in the 1980s," Dale says. "I realized that nothing is forever, that there's nothing sacred out here. If the right guy comes along with the right plan and puts up the right money, these lands are history.

"The things we do in this world affect others, even at teeny weeny Wheaton's Lodge. My ability to retain a steady clientele affects the livelihoods of my kitchen staff and my guides. My ethics with respect to the fisheries affect the people who use these lakes. Whenever you do anything, there are consequences which are borne publicly and which are borne privately. The larger you are as an entity, the more responsibility you must bear."

$\mathcal{D}$ale beaches his canoe on the southeastern shore of Muncie Cove, a long, narrow finger of Spednik Lake. He sets the picnic hamper on a makeshift table made of planks nailed to a cluster of fir trees and pulls out ham steaks, raw potatoes and onions, homemade bread, pie, and coffee. Within minutes, he has a fire blazing and water boiling and he's cutting potatoes and onions into a black iron skillet. We slap at mosquitoes and look out over the water, talking about the morning's fish. Dale sets out plates and loads them high with food. Everyone eats two helpings of everything, ending with thick wedges of blueberry pie and strong coffee. Dale relights his cigar.

By the time we're finished with lunch, a light riffle that Dale calls a "salmon chop" has risen on the water. He goes to the canoe and returns with a handful of bright pink and blue salmon flies and lays them carefully on the table. Salmon, he says, are less reliable than bass, so he takes them much more seriously. "Some days, you can't raise a fish. Others, you can't believe how dumb they are." As he talks, he picks the flies up one by one and inspects them closely.

The land we're sitting on, Dale explains, is part of the conservation easement that the Baskahegan Company is selling to the state. The easement prevents subdivision and development 500 feet back from the water's edge, while allowing Baskahegan to continue to cut timber. Since state laws already restrict logging right along the shoreline, and because he trusts Chuck Gadzik's forestry sense, Dale thinks the easement is a good solution. He particularly likes the fact that the land will remain in private hands.

"I have a lot of confidence in the private sector's ability to accomplish things," he says. "I think you have to create a world in which this can remain in private ownership, managed in a way that is environmentally sound, and yet still generate a profit. I don't believe we'd be better off in a world of public ownership. You don't have to be a private enterprise to exercise abuses and mismanagement. The public sector can screw up just as effectively as the private sector can."

Dale thinks that most of the north Maine woods should remain privately owned, but subject to restrictions that limit abusive logging and intensive development. In return, he favors incentives and tax breaks for landowners so they can make a living growing trees rather than resorting to development. When asked to define how land-use restrictions should be written and enforced, he describes a common concern: that the rules will be too restrictive and unrealistic if imposed solely from outside by state or federal agencies; but too lax and unenforceable if imposed solely from inside, by small towns.

"Like any other businessman, I get pecked at in every direction from government," Dale says. "They dream up more regulations and restraints and things to tax you on and papers to fill out than you can conceivably imagine. Even a little numb nut like I am, here in the woods, is constantly doing stuff just because it's part of the regulatory environment that I live in."

At the same time, Dale says, as long as property taxes are a town's bread and butter, the need for tax revenues will always undermine the effectiveness of local regulations. Maine state law requires towns to adopt minimum shoreline zoning ordinances, but the promise of added tax revenues from a new lakefront subdivision often persuades towns to apply the least-restrictive rules. By the time towns realize that servicing such new developments often costs more than the additional tax income, it's usually too late. Local residents, he concludes, simply can't regulate their neighbors.

"There has to be a new set of rules," Dale says. "In order to protect what's left, somewhere somebody in authority has to have the balls to say, 'This is it. There shall be no more subdivisions in this area. However, since that is an encroachment on private property we are going to be fair. We want you to operate at a profit, manage these lands as best as you can, generate some capital, and we'll try to revise the regulatory and tax environment so you can survive under this basis.'"

Dale picks up a fly. He smooths the feathers and sights down the two-inch shaft of the hook. "My objective," he says, frowning at the fly, "is to retain the undeveloped status of land. I don't care who owns it as long as it stays undeveloped."

Dale bends the hook slightly, sights along its length again, lays it down. Finally he rises and gathers the pile of flies in his hand. He flicks the butt of his cigar into the dying fire and douses it with the remains of the coffee.

11

Continuum

On a bright, blustery October day, in the northern part of the Baskahegan Company's ownership, Chuck Gadzik follows Dana Marble down a two-track skid road. Big spruce, yellow birch, and fir trees sway in the wind, their branches squeaking and groaning. From the road, narrow trails angle off and disappear into a cool, mossy gloom, penetrated here and there with shafts of sunlight slanting through the canopy.

Dana, who has logged the Maine woods for sixteen years, is in his forties, with thinning brown hair and a Lincoln beard. As he walks, he kicks the forest litter aside, looking for mushrooms. He has just finished logging this stand with horses, and has gotten used to finding chanterelles and other edibles. He turns up one of the narrow trails and stops next to a tall spruce tree.

"I had never logged with horses for a living before, so I had a lot to learn," Dana says, his voice resonating in the deadened echo unique to deep woods. He touches the spruce's scaly bark and looks up into the tangle of branches. "A lot of thought goes into dropping this wood. Like this one right here. When this goes down, it's going to lodge between those two smaller trees over there. We'll put the horse on, and he'll pull the butt of the tree out until it falls. Then you cut your log out, put the reins up on his back, and give him a little slap. That Bud horse, he'll just pull the log right down this twitch trail and out to the skid road."

Dana looks at Chuck, his dark eyes probing.

"This is pretty much what you expected, isn't it?"

149

"Oh yeah, it looks great, Dana," Chuck replies quietly. "I never had any doubts about the results on the ground."

Dana relaxes. "We weren't set up to be as productive as we could be, but the basic concept works beautiful. The old timers did it, so we should be able to. It's just a matter of getting everything to work."

Dana started this job four weeks ago, after months of preparation. He bought two horses, some harnesses, and a van, and he recruited two men who had experience logging with horses. Promising he could do a better thinning job than the sophisticated Valmet logging machines that Baskahegan often uses, he talked Chuck into a trial run. Dana had years of experience logging with a skidder on land owned by Georgia-Pacific, and he expected the operation to go smoothly.

It didn't. Dana and his crew lost one day when their truck lost a wheel, and another when a horse threw a shoe. One woodcutter never showed up, and another broke his hand in the reins. They had to haul the animals in and out every day, burning precious hours and fuel. Only one of the horses – Bud, the one Dana had rescued from a glue factory – would pull big logs. It was hard maneuvering where the hardwood trees were closely packed together, and the loggers were forced to detour into the larger spruce stands. In four weeks, the three men cut 80 cords of wood – less than half the volume Dana could have harvested with a chainsaw and a skidder, and one-sixth the Valmet's output. With expenses outpacing what they would be paid for the wood, they had to quit.

Dana says he's undeterred. He wants to set up a nonprofit logging cooperative that will provide woodcutters with an hourly rate, training, insurance, and other benefits. To train young woodcutters, he hopes to set up an apprenticeship program with retired horse loggers as the masters. He has already invested $13,000 in equipment; to make the cooperative work, he needs another $30,000 for rigging, two new animals, and a portable stable so he won't have to truck in the horses every day. He points out that $30,000 is half the price of a large skidder, but he can't borrow against the cooperative because it doesn't yet exist. And he isn't ready to mortgage his own conventional logging equipment.

"I'd rather cut this way than the way I've been forced to cut on G-P, although they have improved," Dana says. "You can see more of a future here. I'll probably make a lot less money, but if I can see some future, I'd rather work here. Even the Valmet I don't think does as a good a job as this. You don't see too many barked trees, do you?"

"Nope, I don't," Chuck says. "Have you hauled any wood?"

"Yeah, we hauled three loads. One of spruce logs and two hardwood pulp."

"Okay, we'll get you squared up for those." Chuck pauses. "What are you going to do now?"

"I'm going back on G-P for the winter as Northeast Logging. I want to come back here, Chuck. But I'm not going to, until I am really set up to do it right."

The Baskahegan Company owns a piece of land along Route 6 west of Topsfield where Chuck has been culling the lower-grade hardwood and fir trees and thinning the spruce. He's trying, as usual, to accomplish several goals at once: provide room for the biggest trees to fatten into high-quality sawlogs, leave some mid-sized trees to grow into the overstory, and retain the canopy to protect the thick spruce regeneration on the ground. He enjoys walking through the stand in the late afternoon, when the setting sun softens the gray trunks and highlights the seedlings underfoot.

In this location, Baskahegan's land surrounds on three sides a parcel that the log broker H.C. Haynes bought several years ago and stripped of virtually all the trees. When Chuck crosses over onto Haynes's property, he finds that much of the land is now covered with raspberry brambles and pin cherries, the brushy invaders that Baskahegan's modified shelterwood cuts and thinnings are designed to suppress. Yet he also finds pockets on the Haynes land that are filling with vibrant young spruce saplings, where seedlings had been established before the forest was cut down. Recovery in the aftermath of such rough treatment always startles and pleases him. He says he has seen similar results in heavy cuts on Georgia-Pacific land, and on Baskahegan's own holdings that were hit hard by the Dead River Company.

"The resilience of this forest is just overwhelming," Chuck says. "The land is continually presenting options. For whatever reasons, sometimes we ignore those options and do the worst possible thing. But it's just a matter of time before another wave of options comes along. We don't control the forest. It's something you can run with or fight. But you can't beat it." What Haynes has done, he says, "is lower on the continuum than what we are trying to achieve, but it's not a complete failure. And I'm sure beyond us is another, better level."

The harder you fight the forest, Chuck believes, the longer you have to wait for the next wave of options. That was the mistake John Kelley and the Dead River Company made, and the results left the Baskahegan Company in debt and the land impoverished. To Chuck, running with rather than fighting the forest means encouraging the diversity that nature provides, cultivating small stands according to their natural boundaries, cutting more often but more lightly, spreading the harvest pressure over the entire holding, and constantly trying to increase the land's value by improving the quality of the growing stock.

Baskahegan's current average annual harvest of 35,000 cords is approximately 75 percent of the total amount of wood that grows on the company's land each year, a cutting level that keeps the timber budget well in the black. Most important, Chuck leaves the more promising trees to mature as long as possible. His harvest of premium white pine and spruce trees is less than 10 percent of growth, whereas in the thick tangles of scrawny hardwoods, it is often over 100 percent; thus the harvesting pressure is concentrated on the lower-value wood, leaving the higher-value trees to grow. The Millikens are so confident that this approach makes long-term economic sense that they are buying 10,000 acres of cutover Haynes land. They plan to nurture the land back to health, even though these stands won't produce logs of any real value for at least thirty years.

It is this emphasis on long-term goals, common in forestry theory but rare in practice, that attracted Dana Marble to the Baskahegan Company. The fit was close enough that Chuck was willing to offer the horse logger a better-than-average stand of wood and a generous per-cord rate. It soon became clear, however, that Marble could make money only if he got to cut Baskahegan's very finest stands. Logging with horses, therefore, would ultimately lead to high-grading. Chuck thinks that Marble has a great idea about forming a woodcutters' cooperative that could improve the loggers' income and benefits, but he questions whether the cooperative should emphasize horse logging.

"Dana agrees with the critics who argue we should go back to a time when there was much more labor in the woods," says Chuck. "That would be nice. There would be more dollars going for wages instead of to banks and equipment manufacturers, there's no question. But he needs to make a hard reality assessment of what the production and costs really are.

"I asked Dana, 'How many twenty-two-year-olds do you know who, given the choice between sitting in an air-conditioned cab or getting up close and personal with a horse, would choose the horse? Probably ninety-nine percent wouldn't.' He said, 'Yeah, but I want that one percent.' The truth is, you can't build an industry on that one percent. The future is not made by going backwards."

On the other end of the spectrum, Chuck is also unconvinced that the future lies in more high-yield, intensive management. He thinks it's foolish to clearcut a forest, and then bring the trees back with planting and herbicides, when there are so many more less intrusive and less expensive ways to encourage natural regeneration.

The paper companies plant certain species as a strategic hedge to protect their mills against higher wood prices in the future, but Chuck is certain the plantations will cost the companies more than they are worth.

In this preference for working with natural diversity, Chuck is philosophically aligned with many environmentalists who criticize industrial forestry, and he supports the basic tenets of new forestry and ecosystem management. But Chuck believes that forestry needs adjusting, not overhauling, and that the adjustments should be more economic than political.

"The quality of forestry is discussed as a socio-political issue, but it's really an economic one," he says. "The more money we get for the product, the more accommodating we can be of other interests," including recreation, aesthetics, and wildlife. Landowners, Chuck says, can afford to manage their forests with a lighter hand only when they can market all kinds of wood products, from the lowest-grade pulp to the most valuable birds'-eye maple. Like species in an ecosystem, each component of the wood market is critical, and each is connected to the others: If too many pieces are lost, the whole system collapses; if one becomes too dominant, the same happens. While critics complain that whole-tree harvesters have displaced some hand crews and virtually all horse loggers, the machines are the least-expensive way to harvest low-value wood. Harvesting low-value wood, in turn, is what allows Baskahegan to stay in business without high-grading, to pay hand crews a bonus to take extra care, and to use the Valmets in the highest grade stands.

"The key is diversity, diversity, diversity," he says. "Without a healthy manufacturing base, you just won't have healthy forests. Without healthy forests, there's no reason to hold onto forestland. That's when land will get sold out of the industry and developed."

One night during dinner at The Log Cabin Restaurant in Topsfield, Roger Milliken, Jr., Chuck's boss, told a story about the time he toured Baskahegan's land with students of an environmental ethics class from Bates College.

Roger took the students to one of the company's best spruce stands to watch the Valmet operate, figuring they would be impressed by how carefully the machine could weave through the woods and pick out individual trees. Problem was, several of the students had never seen a live tree being cut down. Horrified at the Valmet's crunching power, they started challenging Roger about living more lightly on the earth. Roger suggested that the students think of the entire forest rather than individual trees, and he showed them how the thinning was allowing sunlight to reach the young saplings underneath. But he wasn't getting through to them. Finally one young man spoke up, admitting that the logging job looked much better than others he had seen. Another student snapped back, "Yeah, and if you put a drunk driver next to an ax-murderer, the drunk driver probably looks pretty good. But that doesn't make it right."

"What I had to say couldn't penetrate the shock of watching living trees get cut," Roger told us. "I think that's a real fundamental problem. I talked about the houses they were living in and the books they were reading, and they responded with a simplistic notion that if you lived more lightly, that stuff wouldn't be necessary. There's an attitude that somehow it's wrong to cut trees."

Roger and Chuck are among the most politically active officials in Maine's forest products industry. Roger is a director of the Natural Resources Council of Maine, a prominent environmental organization, and chairman of the Maine Forest Products Council, the state's primary timber trade association. He was a principal proponent of the Maine Forest Practices Act of 1989. Chuck is active in the Society of American Foresters, the Maine Environmental Priorities Steering Committee, and the Baxter Park Scientific Forest Management Area advisory committee. He served for five years, including two-and-a-half as chairman, on the Land Use Regulation Commission. Roger and Chuck both participated in the Northern Forest Lands Council, and their opinions are sought and respected by lawmakers, conservationists, and industry leaders.

Given Chuck's practical experience in the woods, some conservationists would like him to define more clearly where the proper balance lies in the continuum between horse logging and cut-and-run tree mining – between, if you will, Dana Marble and H.C. Haynes. Here he is, these critics say, working with the natural flow of the forest in precisely the ways that proponents of new forestry are advocating, yet he refuses to be openly critical of less enlightened practices. "He should be out front, but he has this freaking stubborn streak in him," one prominent Maine environmentalist said. "He qualifies everything right off the map, until it's meaningless."

Chuck's reluctance to criticize the work of other landowners and foresters has complex roots. Because of the Millikens' wealth and philosophy, he works under less-demanding income pressures than most other foresters do. He truly believes in the natural resilience of the forest, and he knows that Baskahegan's land is in a region with particularly good growing conditions. Significantly, he does not plan to stay with Baskahegan forever, and is doubtless hesitant to anger prospective employers. His tenure at the Land Use Regulation Commission, moreover, made him leery of broad-brush policies that can unduly restrict the majority of responsible landowners in the name of preventing abuses by the few. And he correctly points out that Baskahegan is not unique; many other nonindustrial landowners, because they don't have mills to feed, have similarly longer-term views of land management.

But there is another, more subtle reason that Chuck doesn't push his brand of forestry as aggressively as some environmentalists would like. He is frustrated by anti-logging sentiment such as that expressed by the Bates College students, and he doesn't believe that environmentalists work hard enough to dispel such simplistic rhetoric. If critics want foresters to be more candid about poor logging, Chuck thinks the critics should work harder to help solve the pressures that drive such practices.

In the manufacturing sector, for example, Maine needs more local mills to fashion hardwood and cedar trees into solid wood products such as I-beams, trusses, and furniture parts. In land ownership, Chuck would like more non-traditional investors to follow the lead of John Hancock Timber, which purchased 138,000 acres in Maine in 1993, and has publicly committed itself to long-term, quality forest management. In logging, he hopes more contractors will invest in sophisticated equipment like the Valmets, which have both economic and environmental benefits.

Chuck believes that filling in these missing pieces will lead to better forestry and healthier forests, but only if environmentalists take care not to discourage potential investors. For this reason, he was disappointed in 1993 when several environmental groups in Maine joined a legal intervention of the Federal Energy Regulatory Commission's relicensing of paper company dams in Millinocket, Maine, and several other mill towns. While there were legitimate issues to address, Chuck says, the environmental groups did not fully explain their position to local people, nor did they consider the tone of the anti-industry message they were sending. "The people who make choices about investing are very open and direct about what they consider," Chuck says, "and issues like these directly impact their decisions."

Sometimes his frustration is more personal. In 1994, Baskahegan joined Georgia-Pacific and Champion International in an Atlantic salmon restoration program called Project SHARE (Salmon Habitat and River Enhancement). Designed to help bring back New England's wild salmon stocks by improving the management of key rivers in east central Maine, the project was endorsed by the U.S. Fish and Wildlife Service and several fish conservation groups. The previous fall, a small Massachusetts organization called Restore: The North Woods had filed a petition to list Atlantic salmon under the federal Endangered Species Act. Restore staffer David

Carle shrugged off the cooperative effort in Maine as Project "SHAM." It was a good sound bite, but Carle's attitude unnecessarily insulted one of the conservation community's best potential allies in the forest industry.

In the increasingly tense debates over the nation's woodlands, foresters are caught at the intersection between ecology and economy. As a result, they often feel pressed to choose sides between the environmental community and industry – a painful decision, given that most people go into forestry because of their passion for the environment. Many foresters, including Chuck Gadzik, recognize the industry's shortcomings, and they are enthusiastic about the potential for such concepts as new forestry and ecosystem management. But when they talk publicly about the need for change, their optimism often creates expectations that the shifts can take place overnight. Then, when forced by markets or taxes or quarterly dividends or weather or any number of other factors to cut harder than they would like, they are cast as hypocrites.

These are Chuck's thoughts as we hike up Musquash Mountain, talking quietly in the cool May air. It is evening, and the rose sky is streaked with wispy white mare's tails. Chuck walks with his head down, his voice subdued.

"Have you seen the articles from the west coast?" he asks. "Some people are framing houses out of aluminum now instead of wood. There are probably doctors and lawyers proudly displaying their aluminum studs and saying, 'Aren't I wonderful for not using wood?' Talk about an aberration of an environmental ethic. I guess they've never been to an aluminum mine or a smelter. That just seems so hopelessly naive."

In the face of such misplaced priorities, Chuck says, foresters often retreat.

"Foresters are realizing they need to respond to more and more concerns," Chuck says. "But lots of them get burned, and then they just put up a shell. Like if some paper company forester gets up in public and says, 'Yeah, I made a mistake on that cut, and I wouldn't do it that way now,' he learns not to be that frank again. Because it will come back to haunt him. So what happens is we get out of the debate. We don't participate. I don't know what to do about that."

We crest the summit of the mountain, and rest for a minute at the foot of a steel lookout tower. Then we climb up five flights of stairs and stand on the corrugated decking.

From here, it's impossible to tell where Baskahegan's land starts and ends – the entire ownership is lost in an immense sweep of green and brown and blue. Chuck brightens as he points out the Penobscot and St. Croix watersheds, Baskahegan and Spednik Lakes. Farrow Mountain rises just to the east, Katahdin far to the west. We try to find the fields around his house, but the details are blurred in the endless bumps and dips and curves of the earth. All is soft and broad and open.

Chuck leans on the railing, looking out at the land. The wind pounds into us. It hums through the struts of the tower and whistles through the guy wires. Chuck holds onto the rail. There are two more floors above us, but this is as high as he likes to go.

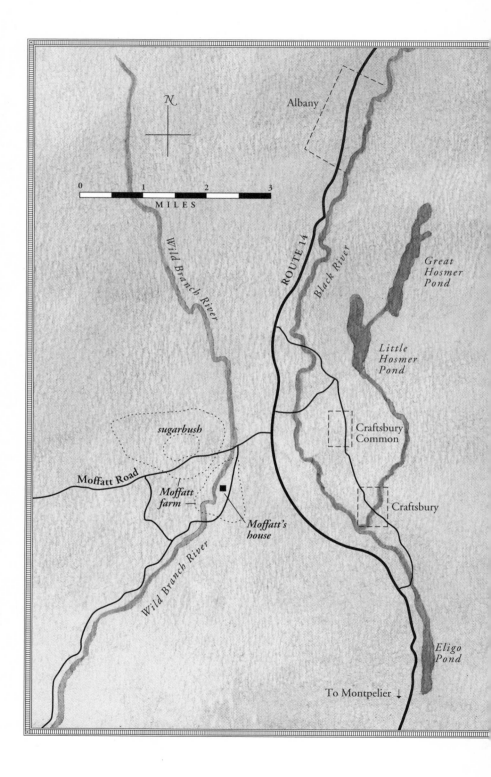

N

0　1　2　3
MILES

Albany

Wild Branch River

ROUTE 14

Black River

Great Hosmer Pond

Little Hosmer Pond

sugarbush

Moffatt Road

Moffatt farm

Moffatt's house

Craftsbury Common

Craftsbury

Wild Branch River

Eligo Pond

To Montpelier ↓

Part Three

VERMONT

—

*". . . they said that the weather and seasons
that form a land form also the inner fortunes of men
in their generations and are passed on to their children
and are not so easily come by otherwise."*

CORMAC MCCARTHY

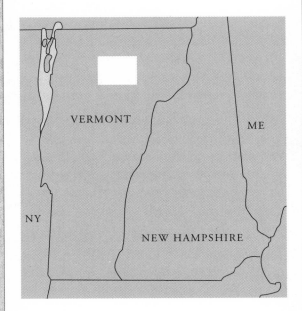

VERMONT

ME

NY

NEW HAMPSHIRE

*Caspian
Lake*

AREA OF DETAIL

12

Cutting Maple

When he's twitching logs through the woods, Jim Moffatt sits up alert in the seat of his skidder, checking over his shoulders to make sure he doesn't smash or crack or scrape any standing trees. When he's just driving his skidder down his woods road, he slouches a bit, round-shouldered, his mittened hands resting on the wheel. As he drives, he watches for deer. The deer winter in the valley bottom, in thick stands of fir that shelter them from the snow. In the mornings they walk the packed snow of the dirt road – Moffatt Road, named after Jim's father – up the hill to Jim's log landing, then through the landing and along this packed skid road to nibble at the hardwood saplings in his sugar bush. Most mornings when Jim drives the skidder in, he pushes several deer up the road in front of him, further into the bush. They move slowly until he gets too close, then leap off the trail and into the deep snow among the maples.

Jim stops the skidder and points at the snow. "Here you see," he says, "a rather cruel drama." In the snow next to the skidder are two sets of animal tracks, one a series of deep holes, the other a shallower set running alongside.

"A coyote chasing a deer," he explains. "You can see the deer plunges right through, but the coyote stays on top. I'm all for the coyote, because we need a predator up here. But still. That's not pretty to see. The deer are having an awfully hard time in this snow."

He puts the skidder back in gear and drives toward the morning sun. Through the screen of the empty branches he can see east across the valley of the Black River to the snow-covered farm fields

163

leading up to Craftsbury Common, three miles away. Then the road curves left, and he turns his attention to his driving. Dropped tops, wood chips, and small mounds of sawdust mark the areas where he and his son Steve have been working. From over a small rise comes the sound of a chainsaw. Steve is felling a maple. Jim tops the rise and stops to watch. Steve is half-hidden behind the tree he's felling until the tree starts to give way, and then he steps back, looking up at the crown to make sure it's going where he wants it to. There's a tearing sound as the crown pulls loose from the canopy, then an airy rush as the tree, a maple almost two feet across and sixty feet tall, accelerates toward earth. It hits with a dead thump. Then the woods are quiet.

Jim, on his skidder, sighs and shakes his head.

"I cross my fingers every time," he says quietly, lest Steve hear. "You teach him everything you can, but what it comes down to is you just have to hope he never gets hurt."

With that he restarts the skidder and drives down to the cutting site. Steve stands at the edge of the skidder road. He smiles up at his father. At twenty-six, Steve is lean and strong. Though Jim wouldn't let him work a chainsaw until he was fifteen ("You've got all your life to learn how to cut your leg off," he told him), Steve began doing woods chores years before. He started off shoveling manure dropped by the woods horses they used to own, then stacked firewood for the woodstoves in the house. Then he started bucking the sugar wood into 4-foot lengths, pruning and harvesting Christmas trees, tapping the maples in spring, and finally, felling trees. When Steve went to the University of Vermont, where he studied plant and soil science, he'd return summers and cut wood to sell for tuition money. After graduating, he worked a short stint as assistant arborist for the city of Burlington, an hour's drive west. He liked the job, but lost it when budget cuts eliminated the position in 1991. "I got axed," he says. He called Jim. Jim said, "Well, you could always come back and sugar." Now Steve works almost every day with his dad, testing the idea of taking the place

164

over in a few years. Seven hundred acres of maples and fir. A hundred thousand Christmas trees. And the "freedom," as Jim puts it, "to work as many hours as you can stand."

Steve likes the work. Give him an excuse to hurry, such as when his father needs a hand with hitching a log to the skidder, and he'll actually *run* through these woods, even in deep snow. He blasts through by leaning forward and plunging his knees through the snow as he pumps his arms, elbows out wide for balance. The snow flies around him. He seems to run not from any frantic sense of hurry, but from the pleasure of being able to plow through snow even after working hard in the woods for hours – a joyful, youthful run. Perhaps he is grandstanding just a bit, to show his dad that he can do this, or that he's *willing* to. But he seems to run mainly for himself. Because he can. To see him run through thigh-deep snow is to *begin* to understand why in temperatures hovering around zero he's wearing only two shirts and no jacket.

Neither father nor son gets cold easily. They keep moving, and they eat a lot. Jim usually brings into the woods a bread bag with three or four sandwiches, some cookies or other sweets, and a thermos of coffee. The bag goes home empty. Jim is known for eating well and enjoying it. He pretends that hunger makes him mean. "Better be ready to share," he'll say should someone eat when he's foodless. When he takes one of his rare short vacations and drives to the Maine coast with his wife, Joan, they stop every two or three hours to eat. If someone removed the road signs between here and the coast of Maine, Jim and Joan could navigate by restaurant. Steve too eats well, particularly when he's working outside. Yet they are both trim. Steve has his father's compact, five-foot-ten body, his square shoulders and arms, the same cool blue eyes sparkling over an alert, friendly face prone to smiles. In the way he talks, the way he moves, and particularly in the deliberate way he approaches his work, Steve resembles his father strikingly. This fact doesn't seem to scare Steve as much as it does many sons his age.

Jim shuts off the skidder's engine.

165

"So what's next, Steve?" he asks from his perch in the skidder's cage. "Where am I supposed to be? You want me cutting, or you far enough ahead I can pull another load out now?"

Jim does this often, handing Steve the decision. It's part of the apprenticeship. Steve bears it well, always working the game for more information. Right now, six logs ahead, he's calling the shots.

"Got a load. Half a dozen logs in a half a dozen places."

"You're going to make me work," Jim says.

"Guess so."

"Where's the worst one at?"

Steve grins, squinting up at the bright sky behind his father. "Worst log? Or worst hitch?"

Jim chuckles. "Worst hitch, I guess."

Steve points north. "End of the road. There's a big maple back behind those ashes."

"Can I get to it?"

"I don't know."

For those who save their wood for good markets, this is a good time to cut, not despite but because of the cold and snow. It is the third coldest and the fourth snowiest winter in Vermont's recorded history. In the six weeks since Christmas, the temperature has climbed into double digits only three times, twice to break into the teens for two- and three-day snow storms, and once into the thirties, when it rained, putting an icy crust on the snow. Most mornings the temperature has been ten or twenty below, breaking zero only after several hours of sunlight. The severe weather has slowed logging crews all over northern New England. It has frozen both engines and men and cut the winter's harvest rate to about half what is normal. The reduced cut has driven hardwood pulp prices from $30 to $40 a cord, "roadside," meaning that if you cut the wood and get it to the roadside, the mills will pay you that and send out the truck as well. Jim has plenty of pulp. He also has a few good sawlog trees, and prices for maple logs are at near-record highs, making some of his best trees worth more than a hundred dollars. Last summer he cut a truckload – about fifty logs, or twenty or twenty-five trees – that fetched him around fifteen hundred dollars. Right now, in this first week in February, he and Steve are pulling out about 8 cords of pulp and at least a couple, sometimes several, high-quality sawlogs each day, grossing a few hundred dollars a day, sometimes more.

Jim is glad to see pulp prices running high, because his woodlot, like most in Vermont, holds relatively few premium sawlog trees. It would hold more if over the decades he and his father had managed their maples for lumber instead of sap. "Sugar" trees – the ones tapped for sap – do not make the best sawlogs, for several reasons. Tapping a tree for sap creates a defect in the trunk beneath the bark – a long, dark stain running up and down the wood from the tap hole. Since each tree is retapped every year, the bottom six or eight feet of these trees, which would otherwise hold the widest, oldest, most valuable wood, are streaked with such defects, and therefore useless as sawlogs. These butts go to the pulp or firewood stacks, along with small, crooked, and gnarled logs.

The other problem, says Jim, is that you manage a maple stand differently for sugar than you do for sawlogs. If you're growing for sawlogs, you let the the trees stay slightly crowded. This forces them to grow tall looking for the sun and keeps their crowns relatively small, thus maximizing the length of the log. In a sugarbush, you give the trees more lateral room, letting the crowns spread, maximizing the foliage to generate more sap.

Jim's entire sugarbush, then, which occupies about 150 acres of his total 700 acres, has been compromised in terms of its value as sawlogs. In addition, the stand that Jim and Steve are cutting today holds some of their poorer trees. The stand is on a steep slope facing east, so the soil is cooler and wetter than in the flatter, better-drained, southeast-facing hilltop above. The trees are shorter, thinner; only a few are big. Not many are even worth tapping. Steve and Jim are cutting down some of the few big trees for sawlogs and most of the smaller, crooked trees for pulp. They're leaving the more promising young trees and a scattering of the best big trees, the straight ones, to serve as seed trees, in the hope that this stand will produce good timber in years to come. As they work each section, they pull the taps and gather the sap lines, for they will sugar here no more. The blue plastic spigots and black polyethylene tubes lie in a confused pile in the snow next to the skid trail.

"Plastic," says Jim, giving a gentle kick to a tangle of tubing. "My father would roll over in his grave."

Jim's father first tapped these trees in 1939. Jim, who is fifty-seven, has sugared here virtually every spring since he was big enough to carry a bucket. "I've probably been to each of these trees about three hundred times," he says. "Tapping, pulling taps, carrying buckets. They look pretty familiar. But this one slope won't be a very good piece of property to own for the next decade or so. We could try to establish another maple patch here, maybe sugar again. It's so ingrained in us to save the maple. But after the way the maple here has grown, we're more inclined to let this stand go as it may. There's some beech trying to establish itself. We may let that take hold instead. Just see what happens." He looks around and sighs. "I don't think *I'll* cut another tree in here, anyway."

It has been several years since Jim's sugaring operation, which dominates two months of his time each year, made any real money, which is part of why he's closing this part of the sugarbush. One reason for the poor return is that a glut of maple syrup, much of it from large farms in Canada, has kept prices low. Another is poor production in his own bush, which Jim believes is caused by maple decline. Maple decline is a mysterious problem affecting maples on tens of thousands of acres throughout New England and Canada. It has inspired multimillion dollar international studies, such as the joint U.S.-Canadian International Maple Decline Project. Despite the studies, foresters and researchers can't even agree whether maple decline is a disease or merely a "condition," a collection of symptoms caused by some conglomeration of unrelated factors such as weather trends or long-term forest successional pressures or bugs or sugar tapping or natural cycles no one even knows about. Jim believes the decline has been caused at least partially by acid rain. On his own land, the failing health of his maples – he has one five-acre stand where virtually every tree has died in the last fifteen years – has been accompanied by the growth of acid-tolerant plants he hadn't seen in his bush before.

The scientists often can't even agree on whether a given piece of forest suffers from maple decline. Jim's sugarbush, for instance, was one of those studied by the Maple Decline Project, and the scientists concluded that his maples showed no signs of decline. To this, Jim shakes his head and points up at the silvery, barkless branches, leafless in summer, that indicate a declining tree.

"Now, I don't want to say anything ridiculous," he says. He laughs, genuinely amused. "I don't want to cast any aspersions. But obviously the complete opposite is true."

Jim has been wanting to cut these trees since last spring, when prices for sawlogs first went to near-record heights, yet things kept intervening. July wasn't as dry as he had hoped; August was dry, but he and Steve were doing a rare outside contract job, cutting oak for Jim's friend Dave Marvin. Autumn was wet, and by the time the ground froze in December, they were too busy cutting and selling Christmas trees to get up here. So they couldn't start until January.

Wood prices are notoriously volatile. They can rise or fall as much as 50 percent in just three or four weeks. Every month Jim waited, he risked losing thousands or even tens of thousands of dollars. The delays worried him a bit. And they made at least one person, his older brother Larry, wonder if Jim had, as Larry put it, "an emotional problem" with the idea of cutting his sugarbush.

"Larry thought I couldn't deal with this because we've worked this sugarbush so long," says Jim. "That it was my father's and all that, and that I worked it as a boy. I suppose this is a minor factor.

But it isn't something that makes me hesitate to the point of not doing the right thing. I'm just not an impulsive person. I can't be hurried into a decision. As a young man, I made decisions that could have waited a while longer, and that hurt me badly. That made me a firm believer: Never commit yourself too soon."

Larry suggested to Jim that he sell the logs as stumpage, allowing an independent logger to buy the rights to harvest these trees, so Jim wouldn't have to cut them himself. "But the exact opposite is the situation," says Jim. "I've stayed with these trees all my working life. I want to be the one that cuts them."

As it happened, his delays have helped. The high log prices held, and January brought the spike in pulp prices. This meant that Jim could profitably take out not just sawlogs but pulpwood, giving him more flexibility about what and where he would cut. It also lets him leave more of the big trees for another time.

Now that he's cutting some of the old maples, he insists he doesn't mind doing it. "I have just one caveat," he says. "I'm not cutting these trees to live on the money now. This money is going to go to our retirement."

Jim often expresses himself emphatically. Still, he seems to say this with more than the usual emotion.

Shovel in right hand, chainsaw in left, Steve walks through thigh-deep snow. With the shovel he slaps the snow in front of him, like a man spanking burgers, to break the crust and save his legs from the brutal ice on top. Weighed down with shovel and chainsaw, he must push hard with his legs to clear a path. A cleared path is critical, for he may need one later for a quick exit. The tree may not fall in the direction he intends, or it might bounce when it hits, racking its butt into the air. On bare ground, Steve might be able to scramble ten or fifteen feet in any direction to avoid such a hazard, but in deep snow, his only escape route is the path he clears, and it's a slow one. His margin of safety is further reduced by the cold. The single-digit temperatures tire and stiffen muscles, or, if these conditions bring on mild hypothermia, reduce alertness.

As Jim says, "You've got to have a head, you've got to use it. You've got to know. You've got to watch. You got to be paying attention *all the time.*" He says this last sentence slowly, with practiced emphasis. He apparently says it a lot. Steve does a flawless imitation, which his father, smiling, looking elsewhere, pretends to ignore.

Slapping snow, Steve makes his way to a large maple. When he reaches the tree he spears his chainsaw into the snow, burying the blade, leaving the engine sticking out so it will stay dry. Then he shovels a trench around the base of the tree. He flings over his shoulder perhaps twenty shovelfuls. Then he stabs the shovel into the crust, pulls the saw out, and ponders how to drop this tree. Normally he likes to leave the tree's butt facing a direct, easy path to the skid trail. But in this case that line of fall would land the tree atop a cluster of promising young ash. Ash is prized furniture material when it gets big, and the Moffatts don't have much of it. So Steve decides to drop the tree to the left instead, perpendicular to the path it will need to follow out. This will mean some tricky skidder work for his dad – the price they'll pay for leaving some ash trees that even Steve may not live to harvest.

Standing next to the maple, Steve points his arm along the intended line of fall. He looks out along his arm, down at the tree, and out along his arm again, trying to see where on the tree he should make his wedge cut. If the tree were perfectly symmetrical, absolutely plumb, and on level ground – if it were an Iowa telephone pole – the correct line of cut would be perpendicular to the intended line of fall. But this tree is none of those. Steve's wedge cut must account for the tree's shape, the slope on which it grows, and the tree's own angle of inclination. Felling is an inexact science, and Steve is still learning. "Some days," he says, "everything drops the way I want it to. Other days I can't hit anything. Or rather, I hit everything."

He pulls the saw's starter cord, squeezes the trigger to speed the chain, bends over, and sets blade to wood. The saw screams, then growls and buzzes as it cuts into the tree. When the middle of the blade is buried, Steve releases the trigger. He looks up to recheck the angle of his cut to the line of fall, then cuts another two inches. Then he removes the saw and makes another cut below his first, angling up at about 35 degrees, to slice a triangular wedge from the side of the trunk. With a twist of the blade he flips the wedge into the snow. The tree now has a mouth, which gapes at its intended line of fall. Steve steps around the tree and begins a cut into the other side. He guides the saw toward the top of the wedge cut. He keeps one eye on the saw and one eye very much on the tree. When the tree creaks and starts to move, he pulls the saw free and starts backing away. The big tree falls ever so slowly at first – the distance between crown and ground seems vast – then suddenly it is moving very fast. Several tons of wood hit the earth. Shards and discs of snow-crust spin through a cloudy wake of powder. You can feel the impact thirty yards away, even through three feet of snow and a pair of thick boots. You can feel it in your collarbones.

The tree lies in a trench of snow. Steve steps onto the trunk and walks toward the crown. Where the trunk turns to branches he steps off and shovels a small clearing in which to work. He yanks his saw to life and cuts through the tree just below where it branches into its crown. Everything below that spot, the bole, will be pulled out by the skidder.

Steve walks back up the hill and gets Jim. With Steve following on foot, Jim drives the skidder down the hill, past the piles of branches and other debris left from last week's cut, and into this corner of the woodlot. Where the skid trail narrows between a young ash on one side and a young maple on the other, he stops, turns the skidder around, backs it between the two young trees, and climbs down.

The big maple Steve has felled lies another 30 feet behind the skidder. If this log lay lined up with a clear path out, Jim would just pull out 30 feet of cable, loop it round the butt of the log, reel the cable in till the butt of the tree met the skidder, and drive off. But the log lies perpendicular to the path out, and that path is blocked by two standing maples, one a foot across and one, only five feet this side of the log, about 14 inches thick. To get the log out, Jim will have to pull it against this larger standing maple and let the log pivot around the maple until it points toward the skidder. Then he'll have to reel the log in without further scraping the 14-inch maple, barking the other big maple, or trashing the saplings that flank the skid trail.

"You got me an easy one," he says to Steve.

"Thought you'd like that," says Steve.

Jim gets in the skidder and backs between the two saplings. Then he gets down, goes to the rear of the skidder, grabs the end of the cable, and pulls it out. The cable is heavy. Jim must lean into the snow to pull it off the spool. Steve follows with a 4-foot-long flexible steel rod with a hook on one end. When they reach the log, Steve pushes the hooked end of the rod through the snow under the log. When the hook emerges near Jim's feet, Jim slips the end link of the cable's hitch chain to it, and Steve pulls rod and chain back under the log. Then he unhooks chain from rod and passes the end of the chain over the log to Jim, who secures it around the butt of the log. They're ready to pull.

Back on the skidder, Jim slowly winches in the cable. The log slides sideways toward the first maple, pushing before it a growing berm of snow. When the log meets the maple, the maple shudders. Jim keeps pulling. The berm of snow pads the log as it slowly pivots around the maple, straightens to the line of the cable, and then starts sliding toward the second maple. When it becomes clear the log will hit the second tree, Jim stops pulling. The log settles into the snow. Jim drives the skidder between and past the two saplings, then reverses, swings the back end left into a patch of open ground next to the skid trail and re-engages the winch. The cable jerks the log forward, pulling it left. It passes cleanly by the maple.

Jim's not out of the woods yet. He must pull the log between the young maple and ash pinching the trail without damaging them. He winches the log in a few more feet, releases the cable, drives back onto the skid trail, then pulls the cable in again. The log, swinging right now as the cable pulls it toward the skidder's new location, splits perfectly the space between the young maple and ash. When the log is flush against the back of the skidder, Jim drives forward until the log is clear of the small trees, stops, and gets down.

"What next?" he asks Steve.

Steve tells him the rest of the logs are up around the curve.

"All right. I'll get those. Then what? You gonna cut some more in here?"

"I don't think so. I think we've cut enough for now."

"Did you look beyond where we were just working there?" Jim says, waving east. "Seems like there were a few further in."

"The other day I looked in there and thought, God there's a lot of stuff in here," says Steve. "Today I don't see much."

"Not a tree within forty miles, huh?" Jim says with a slight grin, looking about at his trees.

"There's that one birch," says Steve. "An ash and a couple maples. Probably best leave them. I don't know. What do you think? How would we get those out?"

"Take 'em out like I just pulled that last one."

"That took twenty minutes."

"I might be faster the second time."

They look at each other a few seconds. Finally Steve says, "I don't know. I think we leave what's here and move back up the hill. We can get some of these junk maple and ash for pulp on the way out, wrap up here, and move up to the main bush next week."

"Okay," says Jim. He climbs into his skidder. "We'll call this patch done."

13

A Farm's History

Jim Moffatt didn't choose his farm's location, but he has never seriously questioned it. The site was chosen by Jim's parents, Bob and Eila Moffatt, and, less directly, by Bob's parents, Joseph Moffatt and his wife Abby, who had come down out of Quebec to Montgomery, Vermont, in the late 1800s. Joseph's family had sailed to Canada from Scotland in the early or mid-1800s – the details are lost to history – and in the following years joined with a French-Canadian family called the Rushfords, formerly the Rouchefords, to form a company to cut and mill spruce clapboards. The Moffatts and the Rushfords would move into an area, establish a logging settlement complete with a small sawmill, spend several years cutting and milling what big spruce they could reach by horse, and then move on to a new area. The two-family business thus moved slowly through southern Quebec during the last half of the 1800s. Around 1890, they moved south across the international border to the spruce forests around Montgomery. By that time, Joseph and Abby, a Rushford daughter, had formed a connubial bond to complement their families' business relations. The bond proved quite fruitful, as over the next twenty years they had seven children.

The youngest of these, born in 1906, was Bob Moffatt, Jim's father. When Bob was about the length of an ax handle, his parents took a logger's retirement: They bought a small farm on Eden Mountain, one ridgeline southeast of Montgomery. They milked cows, grew crops, split shingles – a difficult trade at which Joseph had moonlighted over the years – and raised their kids.

175

Perhaps only a logging family would see a farm as a place to take it easy; Joseph and Abby worked as hard in retirement as most people do in their prime. Eventually they sold the farm to one of their older sons and bought a place on Eden Lake. There they finally took it easy, except when they didn't. When Joseph Moffatt was seventy, his son Bob, finding himself short of hands at haying time, asked Joseph to come to Craftsbury to hay for a couple days. Bob didn't give much thought to making this request of his father, though haying was hard, hot work, and though his father lived seven miles away on the far side of a mountain and had no car. He didn't give much thought to any of this until his father knocked on his door at seven in the morning on haying day. He had walked over the mountain, having started before five. He declared himself ready to hay. From such stock as this did his son Bob, and Bob's sons after him, absorb early the lesson Jim and Larry still cite as the central economic fact of farming and logging: That while you must always work smart, you must often work harder and longer as well.

Bob Moffatt first applied this lesson in the granite sheds of Barre, Vermont, where he moved after graduating from a Burlington business school (a two-year program he entered out of high school) in 1927. The granite sheds were big barnlike structures along the railroad tracks where granite cut from the area's quarries was sawed and chiseled into building blocks, lengths of sidewalk, and headstones. The area's fine and plentiful granite, along with the nationwide economic expansion of the time, had made Barre a boom town. It had grown from two thousand people in 1880 to over eleven thousand in 1910. Most of the men worked in the dozens of quarries in the hills around town or in the granite sheds that lined the Winooski River. Stoneworkers and carvers came from as far away as Italy and Scotland. They cut many of the nation's grave markers, as well as blocks for buildings such as Washington, D.C.'s Union Station, New York City's old Plaza Hotel, Chicago's Cook County Courthouse, and the Vermont Statehouse.

Bob Moffatt got a job running a sandblasting machine, smoothing hunks of granite into flat planes. Dirty work, but it paid well. He saved more money in a few years polishing stone than his father had in a lifetime of woods work. If he hadn't fallen sick from granite dust, his sons might today be cutting granite instead of trees. But after a few years the dust gave him silicosis, an emphysema-like condition common to stone workers. He quit before the condition seriously debilitated him, returned to Eden, and began shopping for a farm.

Bob's timing was good, for while he was working in Barre, the Great Depression had arrived. Farmers were struggling, and land was cheap. The several thousand dollars he had saved was a small fortune. He looked around Eden and the surrounding towns. To the southeast, in Craftsbury, he found a small, seven-cow farm along a stream. It had some bottomland to grow feed grain, some trees for firewood, and was in one of the state's more productive dairy areas.

He bought the farm and had money left over. The work went well, as he found he had a knack and a liking for dairy. He liked pulling milk from the cows, running it through the cooling fins, and putting it up in cans in the big cooling bin. He liked looking over his fields, seeing the grain grow for his herd. But when he married Eila Greene and the young couple started pondering pitter-patter, Bob realized his seven-cow farm would soon prove too small. He and Eila looked around for another. They found a promising one a couple miles down the road going west: a two-bedroom farmhouse, a bigger barn with a silo, a hundred acres, both bottomland and woodland, along the Wild Branch River. A bank had taken the farm in foreclosure and wanted to unload it. So Bob and Eila sold their little farm and bought the bigger one. It was just ten miles southeast of the hills Bob had grown up in.

———

The woods bought by Bob Moffatt in 1932 represented nicely, in their age, diversity, and health, the forest that had taken hold in Vermont over the previous century. It was second-growth forest that rose after timber companies and farmers had stripped over half of the state of its trees between 1763, when the end of the French-Indian War opened Vermont to widespread settlement, and 1850, when the collapse of the sheep industry, the exhaustion of Vermont's soils, and the scarcity of accessible big timber forced farmers and loggers to look west. Photos of Vermont in the 1830s and 1840s show a much barer, more expansive landscape than today's. Even in Montpelier, the gold-domed Vermont Statehouse, which today rises against a backdrop of maple, beech, and white pine, stood before bare hillsides.

As the forest grew back in the late 1800s and early 1900s, the mix of trees changed. Vermont's forest was still dominated by conifers (mainly spruce, fir, and hemlock) and northern hardwoods (primarily maple, beech, and ash); but the hardwoods spread over much more land than they had previously. The newly exposed soils were warmer than they had been, and more favorable to hardwoods than to conifers, and the more numerous, lighter seeds of hardwoods gave them an edge where disturbances had exposed the soil. These advantages helped hardwoods, particularly maple, cover most hillsides in Vermont. The lush, vibrant look of the state today, and the abundance of shimmering deciduous leaves that so pleases tourists in the fall, are partly the product of the clearcuts of two hundred years ago.

Few Vermonters complain about this, for an abundance of maple is one of the state's most attractive distinctions. Further, both maple and the other northern hardwoods produce excellent lumber and first-rate firewood. Maples also make syrup, which has long been important both economically and culturally to the state. And the second-growth forest contains enough spruce and fir that cutting softwoods for lumber is still economically worthwhile.

Along with these forests, Vermont was blessed, in the late nineteenth century, with the presence of major transportation corridors on its eastern and western borders. Since the early 1800s, both Lake Champlain and the Connecticut River had provided shipping lanes for timber, and the railroad, which arrived in 1850, provided another quick way to move goods south and west. The same steam-engine technology that brought the railroads also brought portable, steam-powered sawmills and tractors that enabled loggers to harvest stretches of forest that had escaped the first field-clearing cuts of the previous hundred years. By the 1880s, Burlington, Vermont, situated on the eastern shore of Lake Champlain, was shipping more sawlogs than any U.S. city except Chicago and Albany, New York. Millions of board feet went down the Connecticut River valley as well, either in the river or on rail lines running alongside. Big mills in Burlington and Brattleboro milled logs to ship out of state; hundreds of smaller mills sawed lumber for local use.

For farmers, this trade in wood was too obvious to ignore. Most worked in the woods at least part of the year. Usually they did so in winter, when they had no crops to tend and the snowpack made it relatively easy to draw logs behind draft horses. Farmers without good woodlots cut for mills or timber companies. Farmers with good woodlots could fell their own trees and sell the logs to the mills.

Bob Moffatt wanted to be the latter kind of farmer, and the woods he owned gave him a good basis. But he wanted more. He began buying land whenever he could. He had little cash, but as one of a relatively few solvent farmers in the region, his credit was good. He would borrow to buy a piece of land, then log his new woods to pay off most or all of the loan. At the time, many others, both farmers and loggers, were also buying and cutting land, and the land was so inexpensive they usually sold it afterwards, figuring they could get more land cheaply later. Moffatt kept his. He bought mostly on the broad, gently sloping, southeast-facing hill that lay north of the Wild Branch River – the hill that rose in his back windows, beyond his bottomland farm fields. He bought there because the hill was covered with maples, and he wanted a sugarbush. The maples would give him syrup when syrup markets were good, provide firewood interminably, and would eventually produce good lumber. The softwoods on his wetter slopes and in the cold low spots would give him pulpwood and some sawlogs. In the meantime, the cows would supply the main part of the family's income.

It was a sensible plan: The farm's small scale, manageable by one family, would keep overhead low. The variety of products – milk, sawlogs, pulp, and syrup – would help the Moffatts weather bad periods in any one market. The mix also used the year well: They could log in the winter, sugar in the March mud season, tend young cows, crops, and gardens in the warmer months, and harvest crops in the fall. It was, in essence, a low-overhead, diversified business plan. The trick was in hedging your bets properly and not getting overly invested in any one venture.

Making syrup, then, was part of a calculated plan. It was also a dream, for in a family that loved trees, Bob Moffatt came to especially love the maple. One of Larry Moffatt's early memories is of his father showing him a maple tree, pointing out its distinctive bark, the shape of its leaves, the way it rose to a high crown. His father told him it was a sin to fell a maple unless you'd thought carefully about what you wanted it for and weighed that against what the tree might become. If you were careful not to tap it too heavily, a maple could produce a sweet, salable product while it grew into lumber. And even the worst maples made good firewood.

And they were beautiful.

Jim has a photograph that shows him and his father working the sugarbush. His father stands behind a pair of draft horses. Jim sits astride a small crawler tractor in front of the sugarhouse. The sugarhouse, a shed-roofed shack about twelve by fifteen with a lean-to on one end to hold the sugar wood, was just two years old then; Jim and his father had hammered it together in 1957 to replace the crumbling sugarhouse that Jim's father had built in 1939. Both sugarhouses stood in what Jim calls the Number One sugarbush, which Bob Moffatt had acquired in the 1930s and 1940s.

Sugaring is now a fairly automated process. Plastic tubing connects the tapped trees and carries the sap downhill into collecting tanks, from which it is siphoned into truck-mounted tanks for transport to a large evaporator. Until the 1960s (and at present on a few farms), taps drained into individual buckets. Making syrup required a minimum of two people – at least one to carry the buckets to the sugarhouse, and one to tend the fire beneath the evaporator, which had to be carefully maintained at an intensity that would boil the sap without scalding it. This labor-intensive process necessitated that the sugarhouse be in or near the sugarbush.

The sugaring operation Bob Moffatt established in the 1940s, with over a thousand taps, required three or four people – two or three collecting and replacing buckets, and one managing the fire and adding fresh sap to the boiling pan as the water slowly boiled off as steam. At first, Bob Moffatt and one or two hired men carried the buckets while one of Bob's brothers tended the fire. By the late 1940s, Bob was able to replace the hired men with his children. Eila had given birth to a daughter and three sons during the 1930s: Virginia in 1931, Larry in 1934, Jim in 1937, and Andy in 1939. From their earliest years, the children helped with chores, Virginia in the house and the boys in the barn, woods, and fields. In the one-room schoolhouse the children attended down the Branch Road, their barnlike aroma gave Jim and his brothers no particular distinction. All the farm boys rose early to milk cows and clean horse stalls, went to school, then came straight home to help milk or log or sugar or fix whatever had broken that week. By the time they finished high school, Jim and his brothers knew how to deliver calves, how to boil sap to syrup, how to raise and harvest corn and hay, how to get a pair of draft horses to skid a log out of the woods, and how to fell a tree along a given line of fall without binding and bending the saw or getting themselves crushed.

They knew how to work. They witnessed no life of ease with which to contrast their own. On their way to school each day they saw an embodiment of the strenuous, uprooted life their grandparents had endured, for they passed, just short of the Branch Road's intersection with the main road to town, a large mill operation called the Collins mill. Collinsville, as the mill complex was called, was a tiny company town. Mr. Collins owned several thousand acres of forest, the Collinsville hardwood mill, a commissary, and a dormitory full of itinerant loggers. The loggers, mostly Scandinavians and French Canadians, were hired from a Maine outfit that cut wood all over northern New England and Quebec. Jim remembers going into the forest with his father and watching as the men felled the big maples and spruce with crosscut saws and axes, shouting orders to one another in Swedish and Quebecois French.

The loggers' rough, unpredictable life served as a constant reminder to Bob Moffatt of the consequences of slacking off, of losing his farm. Jim says, "My father believed in going to bed early and getting up early and working long days and being frugal and all that. You could never have a half an hour off to do anything. That would have been sinful! If your hands ever stopped working, you might find time to smoke a cigarette. Which we did anyway."

By 1950, Bob Moffatt owned thirty cows and 700 acres: 200 along the Branch, another 500 on the hill just north. He owned them free and clear: a mix of healthy maple, fir, and spruce, with scatterings of beech, ash, red oak, and birch. Many of his trees were coming into their prime growing years, putting on girth and, in the case of his maples, producing good syrup. His milk was selling well to the expanding New England market. He had no debt. His farm had matured.

His children, too, were maturing, and over the next decade the question of their futures began to press upon the family. The farm obviously presented an opportunity for someone. But their father made it clear that whoever got the farm would not get a free ride. That person would pay for it, somehow. In Bob Moffatt's world you didn't get something for nothing.

Virginia, the oldest, went to college, married, and moved to Pennsylvania. Larry, the next oldest, who had grown to hate dairy farming and chafed under his father's stern command, left for school.

That left Jim next in line. By the time he graduated from high school, in 1956, his parents lacked the cash to send him to college.

"And there wasn't much going on in this neck of the woods," recalls Jim. "That's for sure. When I got out of high school, you had three choices if you couldn't go to college. You could join the service. You could work for a farmer. Or you could work for someone in the woods. Other than that there was nothing. Absolutely nothing. There wasn't even that much in the Burlington area if you wanted to commute an hour and a half. A little farther south, southern Vermont and Massachusetts, the machine tool industry was doing pretty well. But up here: Nothing."

Yet Jim's father could use the help. Even with sixteen-year-old Andy helping part-time, Bob and Jim had their hands full. They were working a sugarbush with almost five thousand taps, cutting pulp and sawlogs, and cutting and selling Christmas trees from the wild spruce and fir that grew around the property.

They were also tending a forty-cow dairy herd, which took more time than ever. In the post-war years, the dairy business grew in both sophistication and scale, and Bob Moffatt grew with it. What had been a low-overhead, low-tech business supplied primarily by small farmers became increasingly automated and dominated by large-volume producers. Driving this transition were genetics and technology: the Holstein cow, the automatic milking machine, and, most significantly, the bulk tank.

It was the bulk tank – a big refrigerated tank that could store hundreds of gallons of milk for days at a time – that really drove the changes. Before the bulk tank, a farmer milked cows by hand, cooled the milk, then poured it into ten-gallon or "forty-quart" milk cans (the kind now sold in antique stores for umbrella stands), which the local dairy picked up daily by wagon or truck. This was a reasonable way to handle the modest flow of milk, some two or three gallons a day, produced by a typical Jersey cow. Jerseys were long popular in Vermont because they handled the hilly terrain well, and because they produced high-fat milk useful for making butter and cheese.

But bulk tanks and milking machines, along with markets that increasingly wanted fluid milk rather than butter and cheese products, made the doe-eyed, fawn-colored Jersey look progressively less attractive (aesthetics aside) than the larger, lumpier, black-and-white Holstein, which produced anywhere from 20 to 50 percent more milk per cow per day. That the Holstein's milk had only half as much fat mattered little, because by the 1950s almost 90 percent of Vermont's milk was being sold as fluid milk. You want to make ice cream, you need Jerseys; you want drinking milk, you go to a Holstein.

With the bulk tank, dairies could pick up milk weekly instead of daily. This efficiency transformed the dairy industry. One estimate holds that if today's milk were still collected and transported in cans, it would cost 40 percent more than it does. The bulk tank thus made milk collected in cans less valuable, since the dairy had to cover the extra cost of collection. Automatic milkers further exacerbated this trend, since the machines could pull more milk out of more cows faster than people could by hand.

But bulk tanks cost two to five thousand dollars, so a farmer needed more than just a couple dozen cows to support the investment in one. The entire trend toward mechanized efficiency and large volumes also hurt the small hill farmer because high-volume production required the high-producing Holsteins, which couldn't forage on rough ground as well as Jerseys did.

These developments meant that as a dairy farmer, you had to grow or get out. Bob Moffatt grew. He took out a loan for $2,500 and bought a bulk tank and milking machine. He began replacing his cows – Jerseys and mixed breeds, or as Jim puts it, "what we just called cows" – with Holsteins. This strategy succeeded for a time. With the postwar expansion and the baby boom underway, the growing Boston and New York milk markets seemed to want no end of Vermont milk. Prices rose and stayed high.

Meanwhile, Jim had taken more than a passing fancy to a young woman named Joan Dyer. He and Joan had been classmates at Craftsbury Academy, the local high school. "We'd dated a few times," says Jim. "But nothing, you know, nothing serious. Things didn't really begin to heat up until she went away to school in Burlington."

By the early 1960s, things heated up to the point where Jim and Joan wanted to marry. Jim felt an increasing need to get established, to have his own stake in the farm. He was receiving a small salary from his father, along with the understanding that he was building sweat equity, but he had no venture he could call his own.

At about this time, Jim's father was feeling more and more the effects of his silicosis. The dusty barn aggravated his inflamed lungs, and getting up early every day drained his strength. Though he was only in his late fifties, he was rapidly becoming too sick to work.

In 1963, he offered Jim the dairy operation.

Jim examined his father's books, talked it over with Joan, and went to the local bank and got a loan. He bought just the cows and the milking equipment at first, with the understanding he would buy the farm in pieces over the next few years.

Almost immediately, he started to go under. Between 1962 and 1965, milk prices, responding to the increased supply created by the industry's technological improvements, dropped to record lows. To make matters worse, prices for feed grain, which usually follow those of milk, rose steadily. This cost was more than incidental, since grain generally accounts for about half the cost of keeping a dairy cow. The two price curves, one falling, one rising, converged, crossed, and parted. Thirty years later, Jim still remembers the damning numbers: $3.75 a hundred weight for milk, $5 a hundred weight for grain.

In the meantime, in 1964, he and Joan had married and moved into the small farmhouse where Jim had been born; Jim's parents moved next door to a larger farmhouse they had bought years before during one of their expansions. Joan worked as a dental technician, but in the mounds of debt she and Jim had already accumulated, her wages made hardly a dent.

"We got further behind every day," says Jim. "By 1965, I was hopelessly in debt. It was the most discouraging thing I ever faced. You're in your late twenties, you're sinking every day. The only way I could compensate was to get up a half-hour earlier every week to try to earn more money away from the farm – logging, sugaring, whatever I could try. But I couldn't pull it out."

Jim and Joan had fallen thousands behind on every account they had: the grain and feed store, the veterinarian, Jim's father, the bank. "We were embarrassed to show our faces at the grain and supply store," says Jim. "We'd go in to buy some more grain or some milk cans or something, and have to ask to put it on the account. It just kept growing and we rarely had any money to pay any of it down. But the owner never said a thing."

"I told him once we really hated doing this," Joan says. "He said he knew we did. He said he didn't say anything about it because he knew how much we hated it, and not to mention it again. Pay him back when we could."

Nevertheless, Jim and Joan felt it was just a matter of time before someone would call in their debts.

They put the cows up for sale. Unfortunately, most other small dairy farms were in similar circumstances, and for six months they got no bites. Finally, a large farm in the nearby town of Irasburg, looking to expand its herd, bought the cows and equipment. Jim and Joan took the cash, gave most of it to the bank, and worked out payment schedules with their other creditors.

"That," says Jim, "was a very depressing day. I felt like the biggest failure in the world. Here I am, I'm recently married, and I'm a failure. The world is falling apart. Very depressing. *Very* depressing.

"The one thing I owned, was by then I'd bought the five hundred acres up in the hills from my father. When I sold the cows, I worked out a deal to get that land and pay him off over the years. I decided I would try to make it with trees. I knew there was potential up there. I knew if I worked hard enough, over a few years I could pay back these other debts I'd piled up. In a way, it was a relief. It was a relief to get the cows gone and start to work in the woods. I'd never doubted my ability to make money there.

"One thing the dairy situation taught me: You have to make sure that it doesn't cost you more to produce something than you can get out of it. With milk, that spread killed me, and I was powerless to do anything about it. But in the woods, I felt I could make it happen. For one thing, we're very frugal people. We keep our own costs down. That means we can make things work where others wouldn't make it."

Jim already owned chainsaws, his father's crawler tractor, draft horses, and a sled. The rest was work. Pulp was selling well, and he had plenty. So he cut pulp, he says, "like one son of a gun." When he needed more wood out, he'd work an hour longer each day, or a day more each week. He wore out chains, blades, tires, his boots, and the horse's shoes. The twelve-hour day, the seven-day week became routine.

Though skidders were in common use by then, Jim used horses to pull the logs out of the woods, because he was good at working the animals and he already owned them. Logging with horses is slower and more strenuous than logging with a skidder. You spend your day moving in rough ground and often through mud or brush or snow. Jim did not mind. In his early thirties, he felt up to the physical demands, and to be able to generate some profit, no matter how modest, inspired him to greater effort.

"It wasn't any different then than before, and it's not any different now: You want to make it in farming, in woods work, you got to be tough. Got to be determined. Got to be *interested.* Got to pull something out to make it work. If you're dedicated enough you can do it. If you're not, you fall by the wayside."

In 1967, Joan gave birth to Steve. From then until Steve went to first grade, she worked outside the home only occasionally. To replace some of the cash income Joan had been making, Jim took advantage of an obvious wage opportunity off the farm. The Vietnam War had taken most of the young men away from the area, leaving building contractors short-handed. At the time, Vermont was beginning the surge in growth that has continued in fits and starts to the present. The state's rural landscape – its clapboarded village commons, its small farms and pastures set among wooded hills – proved alluring both to affluent professionals from southern New England and New York and to young drop-outs looking for what promised to be a simpler life. The immigration of these flatlanders, as native Vermonters often called them, helped create a bustling construction industry. So through most of the summers of the 1970s, Jim banged nails.

Cutting pulp, making syrup, cleaning teeth, banging nails, Jim and Joan climbed out of the hole their cows had dug. They paid off the bank, Jim's father, the veterinarian. In 1972 they paid the last of their debts, to the owner of the grain store. The owner had closed up shop a couple years earlier, for lack of the small dairy farmers who had kept him in business. Jim made that final payment in person, so he could thank the man for his forbearance.

"I knew you'd pay me," the man told him. "I just didn't think it would be so soon."

14

Leafy Hollows and Pig Wallows

Depending on whom you ask, the immigration of large numbers of out-of-staters over the last thirty years can be seen as Vermont's salvation, its destroyer, or both. The former argument is that "newcomers" (anyone not born here) bring needed money, skills, industries, and perspectives. The latter is that new residents are destroying Vermont's rural culture and economy because they impose incompatible urban and suburban standards and values, drive up land prices, and misunderstand and therefore corrupt the culture they are both consuming and changing.

The most obvious effect of this immigration has been population growth. From 1850 to 1960, Vermont's population grew slowly: While the population of the United States expanded almost 800 percent (from 23 million to 179 million) during that period, Vermont's grew only 25 percent, from 314,000 to 390,000. The main reason was that people left in droves. This happened in other parts of the Northern Forest region as well. Young people in particular went elsewhere, seeking opportunity in northeastern cities or on more fertile and easily worked land to the west and south.

But beginning in the 1960s, the rural remoteness of much of northern New England, and of Vermont in particular, proved a strong attraction. Vermont's population grew to 440,000 in that decade (up 14 percent); to 511,000 in the 1970s (up 15 percent); and to 563,000 (up 10 percent) in the 1980s. Having grown 25 percent in eleven decades, the state grew 44 percent in just three more. The town of Craftsbury grew 47 percent during that period, from 674 people in 1960 to 994 in 1990. Both in Craftsbury and statewide, the growth came primarily in the form of young, college-educated people. By 1990, roughly half of Vermont's population had been born elsewhere.

Those who blame these "newcomers" for Vermont's problems tend to stain all with the same brush. Yet the young people who led the immigration actually tended to fall into two fairly distinct groups. One was the back-to-the-land group, which predominated in the late 1960s and throughout the 1970s. The other was a white-collar crowd that came to Vermont mostly in the 1980s to work in professional, high-tech, or entrepreneurial businesses. The difference in the relationships these two groups had to the land is profound.

Bob De Geus, an ex-logger and forester who now works for the state trying to develop innovative forest-based businesses, puts it this way: "Those who moved here in the seventies wanted to as-similate. They wanted to become Vermonters. They wanted to live off the land and work it. But those who moved here in the eight-ies didn't really want to live that way. They wanted to graft. They wanted some of the nice things Vermont has to offer, but they didn't see themselves working on the land. Which is fine. Except that they so often had, and still have, very little understanding of the people who do live that way."

This misunderstanding is worst, De Geus and many others feel, when it comes to forestry and logging. Despite being sur-rounded by woods (Vermont is about 85 percent forested), many Vermonters – even many environmentally minded Vermonters – don't understand the state's forest economy and the vital role it plays in keeping forested areas undeveloped. Most Vermont land-owners are individuals holding fewer than 100 acres, not large cor-porations beholden to stockholders, and most cut their land sparingly, managing for the long term. But these owners must cut *something* in order to pay the taxes and other expenses ownership incurs, and perhaps even to make a few dollars; otherwise, many would have no alternative to selling or developing their land. Thus the most common effect of logging in Vermont is not to destroy the forest, but to maintain its existence. Yet many Vermonters, particularly newer residents, overlook this connection. They may see cows and applaud the dairy farmer's existence; but when they see a logging truck, they feel they are witnessing the forest's destruction.

Such gaps in understanding are not the only dividing line between newcomers and native Vermonters. Divisions most commonly develop over local tax, education, or development issues. Rising property taxes created by expanded services, such as improved police or fire protection or schools, are often seen – and resented – by natives as a newcomer demand. Newcomers and locals often take opposite sides on development issues as well, though which sides each group takes, pro-development or con, often depends on the nature of the development being proposed. Newcomers in small towns – many of whom commute to other towns to work – often fight new development in an effort to protect the quality of life they have found. They are especially likely to oppose an industrial site or a significant subdivision. Meanwhile the locals, though they too may have reservations about the effects of such development, often see precious jobs at stake. Occasionally a particular proposal turns the conflict the other way, with newcomers favoring some project the locals oppose. Either way, the tensions are corrosive: People stop speaking with each other, minor differences turn into major quarrels, and the annual town meetings at which citizens make laws and consider budgets turn into ugly shouting matches. In one half-farm, half-bedroom community near Burlington, the town has split so distinctly that locals and newcomers get their milk, newspapers, videos, and gasoline at two separate general stores on either side of the tiny town green – in a town of only a thousand people.

For the most part, Craftsbury managed to dodge such conflicts through Vermont's years of rapid growth. It helped that so many of the newcomers who came to Craftsbury were of the back-to-the-land variety. Some of these would-be farmers, of course, both in Craftsbury and elsewhere in Vermont, proved utterly incompetent at their new calling, displaying a disorganization and lack of knowledge about agriculture that alternately amused and flabbergasted the natives, as well as a lack of discipline and initiative that doomed them to failure. Most of those didn't last beyond their first winter. Others stayed, however, gaining a toehold by embracing the need to work long, hard, and smart. They devoured not only *Mother Earth News* and *The Whole Earth Catalog*, but farming manuals and agricultural reports. They saw virtue in self-sufficiency, and they wanted desperately to learn the skills necessary to being so.

By chance, Craftsbury had an institution that catered precisely to such desires: Sterling College. Sterling, situated in a group of white-clapboarded buildings just west of the town common, had been a boy's prep school until the early 1970s. In 1973, falling enrollment and a rising interest in the environment sparked by the first Earth Day led the school to change its name from Sterling Academy to Sterling College, and to establish a "grassroots" program designed for college-aged students. The students would work on farms and woodlots in Craftsbury while also studying agriculture or forestry at the University of Vermont in Burlington – a sort of cross between 4-H, Outward Bound, and forestry or ag school.

At first, the new college and the town co-existed uneasily. Relations between the town and the prep school had been cool, and as a nonprofit institution Sterling paid less than the usual tax load, which won it few friends. And many of the students in the new grassroots program, while less aloof than the blueblood Academy students had been, sported long hair, patched-up clothes, and a scent of marijuana, which put some townspeople off. It didn't help that the school was too expensive for most natives of the area. To many of the locals, the upper-middle-class students seemed to be playing farmer, amusing themselves before moving on to college and its subsequent rewards.

The years eased these tensions. Sterling, led by president Bill Manning, a forester, reached out in many ways, such as making extra payments into the town treasury to compensate for the school's low tax rate. More important, the locals saw that most of the students worked hard. The school's hands-on approach to education appealed to the practical, theory-leery Yankees. It didn't hurt that Sterling, though it owned land of its own, often put its students to work on surrounding farms. This let students see how working farmers managed their land, showed the community that the school took its mission seriously, and, not incidentally, supplied the sponsoring farmers free labor and, from the school's faculty, free advice.

Jim Moffatt was among the first farmers to let Sterling students work on his land. With Bill Manning and Jim overlooking their work, the students practiced their logging and woods management skills and learned how to make syrup and prune and harvest Christmas trees. Jim often hired a student or two to help out at Christmas tree season, cutting and dragging trees, or during sugar season to gather sap. Some of the students he worked with in those years stayed in touch for decades. Jim Esden, who went to Sterling in 1975 and 1976 before going on to get a forestry degree at the University of Maine at Orono (where he was a friend of Chuck Gadzik's), returned to Sterling to teach in the early 1990s; he recalls his time on Jim's land fondly. "Here's Jimmy," he says, "known as a pillar of the community, and he's letting this wet-behind-the-ears flatlander work for him. He was real down-to-earth, quiet, clean-cut. And he actually made a living from the land. Which," says Esden, laughing, "was what I wanted to do when I grew up. He was the picture of something I idolized."

For Jim Moffatt, having Esden and other students work on his land was a gratifying experience, for it introduced him to a new and diverse group of people interested in forestry and in making the rural economy work. Unlike many native Vermonters, Jim saw the influx of young people, both the homesteaders interested in living off the land and the young professionals who just wanted to *look* at the land, as a good thing. He scoffs at the notion that some unbridgeable gap exists between "newcomers" – a word he'll take pains to avoid using – and native Vermonters.

"People talk about this all the time," he says, "as if there's an invisible barrier between the two. Which to me is hogwash. If a flatlander is an asshole, they call him a goddamn flatlander. But if he's a regular, reasonable person, no one thinks anything about it. It is what you make of it. The fact is, most of these new people are an asset – in the tax base they help support, in their interest in the school system . . . they contribute a lot. It's true that some demand more services than small towns are used to providing. Schools, police, fire protection and things like that, and *especially* highway maintenance. But I'd hate to have to support a township that excluded people from somewhere else. We'd be a pretty narrow-minded, lopsided community."

The late 1960s and 1970s were an extroverted part of Jim's life, for in addition to his work with Sterling, he held several important positions in town. The first, for which he volunteered, was on a special "pilot board" to help the town consider whether to close its high school, the Craftsbury Academy (which had fewer than a hundred students), and join with other area towns in a new "consolidated" school. At the time, the state's education department was aggressively pushing consolidation, arguing that merging small schools into larger schools was necessary to create consistent standards, broaden educational opportunities, and reduce the per-pupil cost of education. Jim Moffatt, well aware of the tax burdens that schools imposed on poor rural landowners, found the state's arguments convincing. But much of the rest of the community – including most of the old farm families whose members had gone to the Craftsbury Academy – wanted to keep the school. When the town vote was called in 1968, the votes of these consolidation opponents were enough, by a margin of three, to defeat the proposal. Craftsbury Academy thus remained one of a handful of small-town high schools in the state – and has, to everyone's satisfaction, continued to be a fine school. Jim and other one-time proponents of consolidation now feel the town made the right decision. Though expenditures at the school, such as a recent addition, sometimes create grumbling about newcomer demands, the Academy has for the most part unified rather than divided the town. It has given the different families in town a shared stake in their children's future, connected them through school activities, and provided all citizens a common source of pride. As Jim says, "The people of the town knew what they wanted, and they were willing to sacrifice for it. That made it work."

Though Jim opposed many of his neighbors on the issue of consolidation, he did so with enough tact and diplomacy that, when in 1973 a member of the board of selectmen had to prematurely resign his office, Jim was appointed to complete the term. He was subsequently re-elected three times, serving until 1982.

Serving on the board of selectmen is the most difficult civic duty a person can assume in a small town. Officially, select boards, which in most towns are composed of three citizens elected for staggered three-year terms (one member being elected each year, with the most senior member always serving as chairperson), are simply the executive branch of the town government, the legislative branch being the annual town meeting. At town meeting, residents vote to make the laws and approve the budget; the selectmen and selectwomen then administer the laws and budget and do the other things necessary to see that the town runs smoothly. The official duties include hiring any town staff, such as chief of police, administrative assistant, or town planner; creating a town budget for consideration on town meeting day; enforcing town planning ordinances, zoning laws, and other municipal laws; supervising the work of the town clerk and the road agent (both of whom are usually elected); and handling complaints about anything from tax bills to the condition of the roads during mud season. Unofficial duties include mediating disputes between neighbors and serving as general leaders of the town. Perhaps the central fact of being a select board member is that you must make decisions that affect, sometimes profoundly, the people with whom you share your life: those who work for you, teach your children, sell you your groceries, repair your trucks and skidders and chainsaws, buy your wood and Christmas trees, and pass you on the street every day. Inevitably, you make decisions that anger them.

Jim's tenure on the select board spanned the period in which growth was a major factor in Vermont town politics, and often a divisive one. Craftsbury, partially because it grew gradually rather than in convulsive spurts, had avoided the sudden, large-scale subdivision of farmlands that often brought development issues to a head in small towns during the early 1970s. Nevertheless, because many Vermont towns were being overrun by development during that period, several citizens in Craftsbury, Jim among them, argued that the town should develop a plan to prevent poorly planned or irresponsible development. A planning committee was formed in 1973 (Jim served on it for a few months, until he was made a selectman) and, after three years of work, proposed a plan that would establish one-acre minimum lot sizes and a few other modest building and zoning rules. In retrospect, both Jim and others who were on the planning committee realize that the plan they generated was a bit crude, relying too much on prescriptive formulae, such as minimum lot size, instead of incentives or guidelines for directing growth. It was nonetheless a fairly permissive plan, aimed mainly at preventing poorly planned big developments.

However, even before the plan was published, says Bruce Urie, another farmer who was on the planning committee, "the whole effort to create a plan got stigmatized as zoning – the Z word." As the group worked on the plan, resistance built. Jim and the other selectmen, who had no direct say on the plan but who would be responsible for deciding whether to place the issue before the voters at town meeting, came under pressure from some citizens to pull the plug on the committee or simply shelve the report.

Having put the planning committee through three years of work, however, and with growth the major issue in most Vermont towns, the selectmen felt they could hardly keep the proposed plan from public consideration. At a special informational meeting in November 1976 for the committee to present its work, pro-planning and anti-planning advocates passionately argued their positions. There were "plenty of strong feelings," recalls Jim, and a few harsh words, including some grumbling about newcomers. Most of the sentiment ran against the plan, but enough citizens favored planning that the selectmen placed the plan on the ballot the following spring. The anti-zoning contingent prevailed, 111 to 57.

The entire incident had great potential to divide the town along newcomer-native lines, as town plans had in many other towns. And it did, to some extent – most of the people favoring the Craftsbury plan came from elsewhere, and most of the opponents were longer-term residents. In the end, however, the zoning issue did not create lasting divisions in Craftsbury, for several reasons. One reason was that town relations were good to start with, and most Craftsbury residents valued the sense of community and wanted to preserve it. As Bruce Urie says, "We had seen and read about rifts in other towns, and most people seemed to want to chip in and work together – they wanted to see the town operate in the best possible way." It doubtless helped too that the planning advocates included members of two highly respected Craftsbury native farm families – Jim's family had been in town since the 1930s, and the Uries since the 1830s – who worked to bridge the gap between native families and newer ones.

In the end, however, the main reason that the growth issue did not split the town was that planning advocates quickly surrendered once they saw that their neighbors opposed planning. As Bruce Urie says, "In a way, it wasn't a divisive thing simply because no one wanted it. It was clear that you just didn't talk about planning." In other words, Urie, Jim Moffatt, and other planning advocates, faced with the dilemma of how to police one's neighbors on land-use issues, made the perfectly understandable decision to drop the subject. This was probably a wise decision, given the overwhelming opinion against the plan. But it left the problem unaddressed.

By the mid-1990s, Craftsbury had still not resolved the growth issue, nor the related question of how to balance individual landowner rights with larger community concerns. The town had merely sidestepped these problems. In the eighteen years after that first proposed plan, Craftsbury twice confirmed its decision to leave growth unregulated. In the late 1980s, voters had to decide whether to accept funding and other incentives from the state to create a town plan under the provisions of Act 200, Vermont's statewide planning law. A majority voted not to do so. At about the same time, the renewed growth in the state prompted the formation of another planning committee in Craftsbury. That committee completed its work in 1993, producing a plan much milder than the first – "one with no teeth, really," says one of its authors. Yet even that proposal generated so much resistance that even those on the committee saw no use in bringing it up for a vote. Their work was shelved, literally, in the selectmen's office.

Fortunately for the town, most of the landowners who have sold or otherwise converted their land have taken pains to do the right thing. One large farm family, for instance, sold their several hundred acres in the late 1980s to the Vermont Land Trust (a private, nonprofit group), which developed a small part of the property and permanently preserved the rest. And Bruce Urie, whose family has farmed in Craftsbury since 1833, sold conservation easements on his land to the state in 1993.

Unfortunately, the quiet that sits over Craftsbury, as well as its charming landscape and the relative harmony of town relations, could be changed by the first combination of desperate seller and dumb plan that comes along. With no regulations other than the subdivision restrictions provided by Act 250 (which regulates developments involving 10 acres or more), Craftsbury remains vulnerable to both large-scale projects and the slower, more insidious conversion of forest and farm land into poorly planned smaller lots. A single large subdivision in the right – or rather, the wrong – place would almost certainly raise the growth issue in a way that couldn't be ignored, possibly dividing the town. As one citizen says, "A lot of people figure we're all right, because nothing has happened in the last fifteen years. That's not to say it can't happen at any time. I'm afraid it will take something that people really don't want to make people realize that you sometimes have to give up a few individual rights for the good of the community. But the town isn't ready for it yet."

As Jim began his last term on the select board in 1979, it appeared that the town had put its most divisive issues behind it. The Academy was a source of pride to all; Sterling College was fitting into the town gracefully; and the Craftsbury Sports Center, a growing tennis camp and cross-country ski center which was the other obvious potential point of conflict, did a good job of smoothing relations by granting free admission to residents and by sponsoring the Academy's cross-country ski team. The two years since the defeat of the town plan had soothed what hard feelings its consideration had generated. In short, newcomers and natives felt more uniting them than dividing them, and it seemed that Jim would emerge from his tenure unsinged. Then at almost the last minute, he got burned by a conflict arising from a most unlikely source: an annual fiddling contest that had been founded years before by his father.

Bob Moffatt had played fiddle for most of his life. In 1962, he and several other members of the town's Horse and Buggy Club, an offshoot of the United Church, put together the first Craftsbury Old-Time Fiddlers Contest as a way of helping preserve "old-time" fiddling of authentic nineteenth-century tunes – jigs, two-steps, and waltzes with names like "Irish Washerwoman," "Gaspé Reel," and "Pig in the Poke." To the astonishment and eventual dismay of its founders, the contest caught the wave of the 1960s folk revival. By the early 1970s, the annual event was attracting not just serious folk music fans, but thousands of others looking for a weekend of music, beer, pot-smoking, and dancing outdoors. The whole scene began to look more like a rock festival than a fiddler's convocation. Marijuana smoke drifted over the crowd; people got too drunk; some got belligerent. Others failed, when gripped with *amour,* to find a discreet place to express their inclinations.

By 1975, the Horse and Buggy Club had seen enough (this was, after all, a church group), and announced it would sponsor no more contests.

A group of citizens got the idea to move the contest into the countryside and have the town run it. That way, they argued, the common wouldn't be trampled, the hippies could cavort in the bushes, and only those who wanted to would have to see the ruckus. The twenty thousand people attending, tapped for a few dollars each, could provide a revenue bonanza for the town. And there was a perfect spot three miles west of the common: a big field at the intersection of the Branch and Moffatt Roads, a mile east of the Moffatt's house.

Jim, as a selectman, was among the people who thought this could work, and he took on the job of selling the idea to his neighbors. "I convinced them it would be just fine," he says. "That it would be very carefully monitored. Decent. Everything under control."

Recalling it, he shakes his head.

"What a fuckin' sucker I was."

Once moved into the countryside, the contest grew more raucous than ever. People started coming several days early and staying all week, so that the affair became, says Jim, "just a seven-day free-for-all. Just completely and absolutely out of control." It wasn't the pot smoking and free love that caused problems; it was the drunken and aggressive behavior of a few rowdy attendees, most notably some "motorcycle toughs," as Jim calls them – "people just looking for a place where no one was going to bother them, where they could do anything in the world." The things these toughs did at the fiddle contest included camping on, vandalizing, and stealing minor items from some of the surrounding farms; starting fistfights with each other and with people who had no desire to fight; and threatening anyone who objected to their behavior. The county sheriff's department – independent of the town government – let most of this behavior go unchecked. Jim felt terrible for his neighbors. "Their homes are being vandalized, their property being taken, they're being threatened. Boy. It was horrible."

Jim began arguing after the second year that the contest should be ended. There was no way, he warned, that you could have so many people in one place for so long under the influence of drugs and alcohol, with little sleep and no water or proper bedding, and not have real trouble. He failed to convince the contest's supporters, however, who ranged from those who just liked the party to those who liked the income the event brought to the town. "The attitude was that I was the turd-head," says Jim.

Then in 1981, a group of motorcycle toughs turned over a car in one of the parking areas and set it afire. When a crew of the town's volunteer firemen rushed up in the pumper truck to douse the flames, the arsonists took issue. "They didn't just yell at them," says Jim. "They physically threatened them with weapons in a way that you had to take quite seriously. These firemen were big guys, and they came back scared to death."

Two weeks later, at a special town meeting, the town voted to end the contests. "And that was that," says Jim. "The only good thing about it was the timing of the way it fell apart. My dad died in 1976. I thank God he wasn't alive to see what it turned into after that."

Jim says he made enemies through his handling of the fiddle contest, and that may be the case. But when you talk to people around Craftsbury about Jimmy Moffatt, no one mentions the way the fiddle contest turned out. Even those who opposed him on that and other issues use phrases like "the epitome of decency" and "the soul of responsibility." In the long term, Jim's tenure as selectman and his role in the fiddling debacle seem to have cost him no friends. Yet he remembers this as one of the most difficult periods in his life. "I laugh now," he says. "But it was a source of extreme anguish to me for about five years."

When Jim's term on the select board expired in 1982, he didn't run again. He says he felt he'd had his day. "I got to the point I'd rather stay home and read my book," he says. "It was time a younger crowd took over, and they did. Which is how it should work.

"Besides, there are just some issues that are like a tug of war, back and forth and back and forth. I didn't care to get involved in those anymore."

In 1972, Michael and Penny Schmitt drove through Craftsbury Common and saw a house for sale. The Schmitts lived in Manhattan, where Michael worked in investment banking and Penny in advertising. They owned a summer and weekend place in Greensboro, half an hour east of Craftsbury, and had fallen in love with the area's slower pace, the surrounding mountains, and the rolling landscape of farms and forest.

They fell in love too with the Craftsbury farmhouse, an old colonial that needed work but was spacious and basically sound. "We thought it was the most beautiful thing," says Penny Schmitt. They bought it the next day.

Then they went back to New York and decided what to do. At a time when New York City was at the brink of bankruptcy, with garbage heaped in the streets and each day's paper featuring yet more lurid tales of crime, finding the house seemed a sign. Michael and Penny had long talked about making the move to Vermont; the problem was how to earn a living in a place so remote. The old house, they decided, would make a good inn. Opening an inn in Craftsbury was a risky venture, since the town was 3 miles off the main road, was well to the north of the state, and had no major tourist draws. But the Schmitts figured Craftsbury wasn't much farther north than the long-successful ski resort of Stowe, and that with so much of the country turning its attention to Vermont (which was already being written up in national magazines as a refuge from the tumultuous times), Craftsbury's quiet, along with its beautiful landscape, might prove a real attraction.

In the years that followed, the Schmitts made their Inn on the Common a success by almost any standard. They remodeled the main house, bought and remodeled two houses on adjacent lots, stocked the wine cellar with a prize-winning collection of over two hundred wines (they typically keep about three thousand bottles on hand), and hired chefs from afar. They created an inn that is, the *Boston Globe* noted in a 1993 travel story, "unabashedly elegant" – sixteen rooms furnished with antiques, an attentive staff, the wine cellar, and dining that the *Globe* said "rivals Boston's best restaurants."

The Inn's property is at the west end of the broad, flat hilltop on which the Common is centered, so that the grounds drop gently toward the Black River Valley below. The patio, landscaped gardens, swimming pool, and tennis courts look west and north over this valley toward woods and mountains, including Mt. Mansfield, Vermont's tallest peak. Even in the 1990s, the area is relatively untouched by recent development, and tourist traffic is light. Despite the loss of working farms in Craftsbury – there were dozens in the town in 1960, but taxes, competition, and other pressures on farmers had reduced this to about a half dozen by 1990 – the land remains mostly open, and guests may feel they are seeing a town uncorrupted by the changes of the last half-century.

Travel writers have found this vision of Vermont irresistible. The Inn has received good notices not only in the *Globe,* but in many other periodicals, including *Bon Appetit* ("the quintessential

Vermont inn") and *U.S. News and World Report* ("the most elegant of hostels" in "a postcard-perfect Vermont village"), and in virtually every guide book covering New England country inns. These write-ups have attracted a well-heeled clientele, who tend to come from the Boston-New York-Washington corridor in summer and winter, and from farther afield (California and Texas) in the fall leaf season. Nightly room rates, including breakfast and dinner, begin at $200.

The Schmitts know what they're selling. As Penny says, "The fact that the region is so unspoiled is one of the key attractions. It's not like some tourist town where all traces of normal life have been swept away. The town hasn't been junked up or gentrified. You don't raise your eyes and see condos. You see beautiful forests. When the *Boston Globe* wrote us up last March, they said this is the kind of landscape where there really are beautiful red barns and handsome herds of cows to ski by. It's the *appropriateness* of the way people are settled in."

These people, of course, include farmers like Jim Moffatt, whose sugarbush across the Black River Valley provides part of the brilliant foliage at which guests gaze in October. Were he to clearcut, say, a hundred acres of his sugarbush, and sell it to a developer who then put up a dozen homes, the view from the inn would be significantly altered. The few remaining working farms immediately adjacent to the common – the farms that provide those handsome herds of cows for skiers to pass by – play a similar role in this vision, which is as important to the inn as are its wine and food.

Some in town feel the Schmitts have protected this vision too vigorously. The most humorous example of their zealousness occurred in 1980. "The whole thing was pretty amusing, really," says Penny. "What happened was, we built a swimming pool in the back yard for our guests. A very discrete pool with a dark bottom, surrounded by stones. It looks like a pond."

And indeed it does. Yet the Schmitts' nearest neighbors, a Pennsylvania couple named the Houghtons who spent summers in Craftsbury, objected to the pool, which was only a few yards from their adjoining back yard. So the Houghtons bought three pigs, which they put in a pen not fifty feet from the Schmitts's new pool.

This was not, Penny knew, the view of rural Vermont her guests were looking for. She asked the Houghtons to remove the pigs. The Houghtons refused. The Schmitts hired a lawyer, who sent lawyerly letters threatening legal action. As the Houghtons held fast, the pigs became minor folk heroes in town. A local store starting selling tee-shirts with a sketch of a muddy pig under the acronym HOGWASH, which, the back of the shirt explained, stood for Help Our Grunters Win A Safe Home. The dispute made its way onto the Associated Press wire reports.

The case did not make it to court, however, for the Houghtons got rid of the pigs when they returned to Pennsylvania at summer's end. The Schmitts planted a hedge of fast-growing cedar the next spring to provide a privacy screen for both their guests and the Houghtons, and the rancor subsided as the cedar grew. In the end, says Penny, "it was a one-time summer incident. A funny thing that once happened."

Others in town see the clash differently. One person who has differed with the Schmitts on other matters says that although the Schmitts complained about the pigs making noise and bothering guests at the swimming pool, "there was never any thought on their part that the guests swimming at night might disturb the quiet summer evening for the rest of us. Sometimes Penny and Michael forget they have an impact, too."

Penny's response to this suggestion is that while the Inn, along with the Craftsbury Sports Center, in which she and Michael also own a share, does impose certain burdens on her neighbors and on the town in general, the benefits that the business brings the town outweigh these problems. The Inn provides ten full-time jobs, the Sports Center (which features not only skiing but tennis, rowing, and bicycling) about twenty, and the renovation and upkeep of the Inn's three buildings and grounds provide periodic but vital work for the town's contractors. Jim Moffatt, for instance, worked there for a few weeks once in the mid-1970s. Jim's old friend Raymond Reel, who worked in the woods over the years for both Jim's father and Jim himself, laid the stones for the Inn's pool. Raymond's daughter has worked for the Schmitts for fourteen years.

"I respect Penny for what she's done," Jim says. "If you'd have told me twenty years ago that somebody could buy those houses and create a place that outdid every other lodge and inn around and cost two-fifty a night to stay in, I'd have laughed and said Good Luck. But she did it. And she does provide jobs. People say they're not good jobs, making beds or cooking. But you don't see a whole lot else around."

This echoes what many in town say about the Schmitts: You've got to respect what they've done, and they do provide jobs. The "but" you hear is that, as one person put it, "They know what they are selling, and they will fight hard to protect what they see as their interests." Or, as another said, clearly intending the pun, "They're hoggish."

The Schmitts' most notable conflict with neighbors was with the other dominant institution on the common, Sterling College. That story begins with a real-estate deal, winds its way through Vermont's most important land-use legislation, and ends with a legal argument over how much noise clucking chickens make.

In 1985, the Schmitts' fifteen acres were largely surrounded by land owned by a man named Sterling Carpenter, a doctor who lived in Montreal and had inherited the property from one of the old families in town. With land values rising in the mid-1980s, Carpenter decided to sell. The obvious potential buyers were the two adjacent landowners: the Schmitts, whose original inn building and yard abutted Carpenter's land on one side; and Sterling College, which owned much of the land on the other side.

The college bid first, and Carpenter accepted. The Schmitts felt betrayed. They wanted the property badly, as it would give them extra land to extend their gardens and, more important, forever ensure their views. They did not want their inn surrounded by college kids cutting trees and shoveling manure. So they tried to buy the land from Sterling. They offered to fully reimburse the college for its purchase, adding that they would let the school keep a five-acre stretch along the road for a planned library. When that offer was rejected, the Schmitts offered money, the land for the library, and more land elsewhere. Sterling's officers declined this offer too, explaining that they wanted the land they'd already bought because it was contiguous with campus and would make a convenient place to add further facilities.

Such as a barn. Sterling already had one barn, which was leased rent-free from a trustee, but that barn was farther from campus than the proposed site for the new barn. The new barn site would be a short, traffic-free walk along the college's own driveway.

It would also be, to Penny and Michael's horror, in the line of view from their gardens and tennis court toward Mt. Mansfield.

The Schmitts asked Sterling to build the barn elsewhere; the college insisted that the first site was best, and that any other would be too great a compromise.

The Schmitts exercised the only option they had left: They filed a petition to halt the barn under Vermont's Act 250.

Act 250, which was passed in 1970 when southern Vermont towns were being overrun by second-home and condominium development, is the state's primary land-use legislation. The law is unique in the thoroughness with which it reviews proposed developments, for it considers a development's impacts not only on the natural environment, but on neighbors, communities, and scenic quality. Act 250 has come under criticism for being unwieldy (while most permits are granted within ninety days, some appeals have dragged for years), and is a perennial target of attacks by pro-development forces. Yet the law has withstood repeated efforts to gut it, as most Vermonters apparently feel its intent and mechanisms are essentially sound.

To secure an Act 250 permit, property owners or developers must show that their proposed new buildings or development will not unduly affect such environmental and social criteria as water supply and quality, roads, traffic, municipal services, wildlife, natural areas, historic sites, agricultural soils, and scenic quality. Concerned parties, including neighbors, can file petitions objecting to the possible impacts that a development might have on any of these values. A regional commission made up of citizens holds a hearing at which those potentially affected may testify, weighs the evidence, and then either denies or approves the permit, sometimes with conditions; if the permit is denied, the developer can appeal to a larger, nine-citizen state review board. Over 97 percent of all permit applications win approval.

One type of development *not* covered by Act 250 is the construction of agricultural buildings on existing farms. The legislators felt that was a reasonable exception, since protecting farms was one of the principal original motivations for creating Act 250. Sterling bore no clear obligation to file for a permit, but it did file for one, says Ned Houston, a faculty member who had succeeded Bill Manning as the college president, "because we wanted to show we were behind the Act 250 concept." This left Sterling vulnerable to the Schmitts' petition.

The Schmitts hired one of the state's most prominent law firms; Sterling countered by hiring one of the state's *other* most prominent firms; and both sides submitted their evidence. The Schmitts' position was that the barn would compromise both scenic value and, because it was going to be built on land that had always lay vacant, historic values as well.

"You have to understand things at the time," says Penny. "They had a reputation of keeping a truly horrible, filthy barn. You walked by, and I mean it was to throw up. Manure piling up, pigs wallowing, it was a mess. The property was a disaster."

For Sterling's part, says Houston, "We were maintaining that this was an ag site in an ag town, and that barns were once even closer to town. Our barn was below the horizon line from the inn. They could still see Mansfield."

The site was indeed below the horizon line from the inn, and about 800 feet away. The barn's roof, however, would be visible. More important, says Penny Schmitt, the entire barn and barnyard would be visible from the spot where the Schmitts had built a scenic overlook, complete with plantings and a bench, where guests often sat and enjoyed the view of Mt. Mansfield. "If you're on that point," says Penny, "you'd see the whole thing. The absolute serenity behind our tennis court was going to be gone."

The hearing, held in the town hall, was an impassioned affair. The Schmitts' lawyer hoped to establish two truths: that the impact of the barn would be considerable, and that the Sterling College representatives were insensitive, institutional bureaucrats. Sterling's lawyer, anticipating this strategy, told Sterling's representatives to remain calm, matter-of-fact, and to seem vaguely surprised. "We were to approach it like, 'We're just building a barn, what's the big deal?'" recalls Houston.

At the pre-hearing site visit, Houston couldn't resist ribbing the Schmitts a bit. "'Now Michael,' I told Michael, because I know Mike and we get along well enough. 'Which do you think is more out of place historically – our barn or your tennis court?'"

The climax came at the hearing, as the Schmitts' attorney pressed his case as best he could. With Houston in the witness chair, the lawyer decided to bear down on chickens.

"Don't chickens make noise in the morning?" he asked Houston.

Houston said they chucked and clucked a bit when the eggs were collected.

The lawyer asked just how much they chucked and clucked. He wanted to know: Just how many decibels do chickens actually *produce* when they were excited in the morning?

"I told him," recalls Houston, "that once we got them all singing in unison, they were pretty loud; but that usually took a while, so they were usually pretty quiet."

Sterling got its permit, with no attached conditions about singing chickens, and promptly built the barn. The Schmitts re-landscaped their gardens, spending two summers and "a fortune," says Penny, to plant trees, build walls, and lay new walks to direct the view and strolling visitors away from the old point and to a new one facing more northward.

Penny Schmitt says, "This story brings together two things: farming as a sacred cow, and institutional expansion, which in this case masqueraded as a sacred cow. When I lived in Manhattan I was on the planning board in my part of town and my specialty was institutional expansion, and Sterling behaved the same as a big Manhattan institution does. They're dedicated to their mission, and they don't mind who they step on."

Houston and others say that what was at stake "was a vision of the landscape. It was ironic that they objected to a barn on aesthetic grounds. Because that's part of what Craftsbury is – a genuine rural economy, which is what it should be. If you protest about manure and logging trucks, you get rid of those, no one will be able to make a living here. It'll be just a bedroom community. The loss of the genuine land-based economy would be a real tragedy.

"The biggest irony of the whole thing," says Houston, "is that in a painting the Schmitts had commissioned of their view a few years before, the artist, seeming to feel the view was incomplete, added a barn. That barn is in almost exactly the same spot we put ours."

Ross Morgan, a Sterling College faculty member and a consulting forester who works with many Craftsbury landowners, says "Craftsbury's problem is that it's too damned pretty. It attracts people who want to preserve that pastoral scene. But a lot of rural people are far more interested in preserving the rural economy. The rural character they value is much more tolerant about houses falling down and two cars sitting on blocks in the back because they still have some good parts in them. Or someone moving a trailer next to the house that may be unappealing to the eye, but it lets your mother-in-law live next to you. That's illegal in some towns, to have a trailer – you have to send your mother-in-law elsewhere and see her once a month.

"If you zone these people out, if you make it impossible for them to make a living, you end up with a completely gentrified place. In Craftsbury it's still possible, just barely, for a family to support itself through producing milk or wood products and keep their land open. We still have a land-oriented economic base. But you zone them out or make it too expensive, you lose that. You lose that land-oriented economic base."

Penny Schmitt says she's not blind to these connections. She says she understands the link between the farm economy and the landscape that surrounds her. She says many of her guests, particularly those from New England, understand that too.

"But the ones from elsewhere?" she says. "The ones from California or the suburbs somewhere? They don't understand that stuff. All they know is, they don't want to look at ugly fencing and a barnyard full of animal poop."

15

Playing Santa

Jim Moffatt's one hundred thousand Christmas trees grow in several plantations scattered around his farm. He also has a 40-acre plot in Lyndonville, near his brother Larry's house. Today he's working the lot across the Branch Road from his house. The trees trace the slope of the hill in evenly spaced, undulant rows. With the mountains in the background, the trees make a fine Christmasy sight from the Branch Road, particularly when they hold a little snow on their boughs. Jim and Joan sell about a hundred every year to passers-by, most of them local but some of them tourists or people from downstate, who see the cut-your-own sign hanging in front of the house, pull over, knock, borrow a bow-saw, and walk across the road to claim a tree.

Most Jim cuts himself. From Thanksgiving until Christmas he does little else. He works outside for every bit of daylight and then some. In the evenings he and Joan take and make calls to generate orders and see how their wholesale customers are faring. They read and ponder the news, and talk to friends, and try to figure out what the Christmas tree market is doing. From these six weeks, if things go well, they make from half to three-quarters of their year's income.

With a hundred thousand trees planted, Jim says, "the decision about whether to stick with Christmas trees has largely been made." He sometimes worries about having too much riding on this one venture. Moreover, he likes to use a light hand in his woods, and Christmas tree farming is a relatively high-impact form of silviculture. In some of the cleared meadows where he grows

212

his trees – balsam fir, red and Scotch pine, some blue spruce – the desired species regenerate naturally in random patterns from seeds dropped by previous generations, while in others Jim plants the trees in neat rows. Either way, he and Steve must try to control the grasses and hardwood seedlings that sprout from the sun-warmed soil between and around the trees. To do this they either spray with broadleaf herbicides or mow. In some of their upper fields, where balsam fir regenerates readily, they are experimenting with letting the meadow grass grow around the bases of the trees, surrendering the bottom two feet to nature and shaping the tree beginning at knee height. This reduces the need to spray and mow, but requires that someone hand-clip the bottom branches of the tree each year so nutrients aren't wasted on those lower, stunted branches.

No matter what Steve and Jim do about the competing grasses and hardwoods, they must also shear the trees regularly to create the classic Christmas tree shape buyers want. Steve alone spends two or three months during the late spring and early summer just shearing, and Jim and hired people help out for several additional weeks in the fall. Then Jim, Steve, and two or three employees spend several frenetic weeks harvesting and selling between Thanksgiving and Christmas. Christmas trees are a crop, really, though the impact of a field of Christmas trees hardly rivals that of a corn field, with its pounds of pesticides and soil runoff each year. All this intensive intervention means spending yet more time in an area that already dominates the family's labors. Particularly when he starts thinking about Steve running the place alone, Jim begins to wonder whether they've gotten into Christmas trees too deep, just as he was once too deep into cows.

"There is one hell of a lot riding on this," he says. "You're trying to produce most of your income in one short stretch, and if you bungle it you're in trouble. It's awesome what you're trying to do in five or six weeks. You have to take care of customers, make sure they get what they want, that it's good quality and they get it quickly. You have to be right at the going price – if you're a dollar a tree off, the retailers go elsewhere."

Meanwhile Jim must keep tabs on the market. He has to know whether trees are selling briskly or slowly, whether fir or Scotch pine or blue spruce seem to be the favorites this year, whether buyers are feeling flush and buying premiums or feeling pinched and settling for cheaper medium-quality trees. He talks constantly with his wholesale customers to find out what's selling. To the extent he can, he extracts information from other dealers as well.

"The woods are full of information," he says, "but you have to listen carefully to figure out what to cut and where to sell it and be ready for whatever orders might come in. You don't want to be so short on trees you lose sales; but neither do you want to cut too many and have them dry out or end up on your landing on Christmas day. The worst thing you can do is be too cautious. Then you can't fill orders and you lose all your customers. So I try to balance the risk by keeping myself short but always having some way to get more trees quickly. It's intense. Trying to figure it all out is like a six-week, high-stakes poker game."

In the mid-1980s, before the big Canadian tree farms created a glut that drove wholesale prices down from between $12 and $18 a tree to between $6 and $12, the Moffatts sold around ten thousand trees each year. This was ideal, since that is roughly a tenth of their crop, and it takes approximately ten years to grow a good tree. So far this year, Jim and three helpers are cutting, baling, stacking, and shipping about a thousand trees a week, on a pace to sell about six thousand.

The Christmas tree season keeps the whole family working almost every waking hour. Steve is off the farm, selling trees from a retail stand in Burlington, and at night he often drives back to Craftsbury to pick up more trees. Joan has taken a leave of absence from her part-time dental tech job so she can answer the phone and manage the order sheet. She keeps the order sheet – the main record by which the season's business is conducted – in pencil on a manilla file folder opened sideways: a pencilled row for each buyer, a column for each grade and length of each species, with a final column along the right edge for a total. With four species of trees in four grades and four size ranges, the orders can get complicated. The folder's edges are grey and soft with handling, the manilla surface burred from erasures as orders change. The phone rings all day and all evening.

The customers are Boy Scout troops, Lions Clubs, churches, and PTAs; mom-and-pop nurseries and large garden centers; and individual entrepreneurs who rent lot space for the six weeks between Thanksgiving and Christmas. Increasingly, these retail sellers must compete with big discount centers like Home Depot or Wal-Mart. The chains get a low unit cost by buying tens of thousands of trees at a time from big Canadian farms. They sell a "pretty good tree," says Jim, for about twenty dollars. Jim has to spend about twelve to fifteen dollars to grow an equivalent tree, for which his retailers must charge nearly thirty to make any money. This means that these small retailers must peddle not price, but location or loyalty. For this reason, Boy Scout troops are among Jim's favorite customers. He says that scouts – in addition to being trustworthy, thrifty, and courteous – have a built-in, faithful clientele. Because they use volunteers and donated lots, Scout troops can keep overhead low and make a profit on reasonably priced trees. Jim keeps them supplied with good trees, and they come back to him year after year. He likes Lions Clubs for the same reason. Church groups are more iffy. While some are steady customers (one church has been buying his trees for twenty years), many a church group goes into the Christmas tree business hoping to make a killing on a single weekend, and if that weekend brings poor weather or other bad luck, the group can lose its enthusiasm. Every year or two Jim loses a charity or church group to such misgivings, but he usually picks up another one or two to replace it. This year he has a new church in Burlington and a new garden center and nursery in Hartford, Connecticut. He has about thirty customers altogether, ranging from scout troops that buy a few dozen trees to dealers that buy over a thousand and sell them in New York.

From the dozens of orders and cancellations these customers generate, and from guesswork about what the next few days will bring, Jim and Joan decide each night what to cut the next day – not just how many trees, but what species, grades, and sizes. They don't always have the precise mix that buyers want. When they find themselves short on a species or grade (this year, for instance, they're running low on large premium Scotch pine and balsam fir), they'll try to get some from their friend Dave Marvin twenty miles west in Johnson, or drive east to Jim's brother Larry's place in Lyndonville, or if they need to, drive three hours north to one of the big Quebec tree farms. The Moffatts don't make much on the ones they buy, for they can mark them up only a dollar or two. But the trees help them fill orders and keep customers.

Late in the second week of December, Jim works the planted rows across from his house. He walks along the rows of trees, cutting down those that look ready and glancing occasionally toward the Branch Road. One of his retailers, a Vermonter named Larry Demars, is running late. It is almost ten, and he was due at nine to pick up three hundred trees. Demars bought three hundred the week before and, after much deliberation and thinking out loud, requested the same for this week. In the confusion of the long exchange, Jim failed to collect a deposit, a mistake for which he is now torturing himself. As the morning passes he becomes what for Jim is markedly anxious. He looks to the road, then at his watch. Finally he speaks.

"This," he says, "is starting to mess up my day." Demars runs a retail stand on Amsterdam Avenue in New York City, where, Jim says, "anything can happen." It could have happened, for instance, that Demars stumbled across three hundred trees priced so attractively that he decided to forego buying more trees from Jim. This can occur, for example, when some other tree farm's C.O.D. delivery to New York is refused and the truck driver suddenly has a load of trees he's willing to sell cheap.

The other thing that could have happened, the worst possibility, "is somebody could have stuck Demars in the ribs with a knife. *That* would be bad." This too happens occasionally, or at least comes close enough to happening to make Jim slightly leery of the whole New York City tree business. Just the week before, the Vermont man whom Demars had hired to run his stand in New York – we'll call him Al – had been threatened with a knife. There was a bar next to the tree stand, and one night Al saw a fight outside the bar. The next night one of the combatants walked up to Al and stood in front of him. The man said nothing – just stood there – until he was certain that Al saw the knife he held at his side. Then he asked Al if he had seen anything unusual the night before. Al said he didn't know what he meant. The knife man said he wanted to make sure Al hadn't seen anything the night before – any trouble at the bar, any *fights* or anything outside the bar, because if Al had seen anything like that he might feel he should tell the police in case they came around asking, and that, the man said, was something neither one of them wanted to happen. Al said now that he thought of it he was quite sure he hadn't seen anything, anything at *all,* he didn't even remember *being* there. The man with the knife said that was good. Then he left. As soon as he was out of sight, Al called Demars and told him he needed to find someone else to run his Christmas tree stand.

Outright robbery is another problem. In New York City, Christmas trees sell for up to seventy-five dollars. In December, a busy operator might end the week with ten or twenty thousand dollars on hand, sometimes even more. One story tells of a pair of tree stand operators who were closing up shop in the city on Christmas Eve, flush, happy with their season, beginning to feel, as it were, the Christmas spirit, when a sedan pulled up. From each door stepped a man holding an automatic weapon. The stand operators didn't have to ask what was up. They handed over fifty thousand dollars, and were glad to live to tell the tale.

All this rolls around Jim's head as he cuts trees and waits for Demars. He works fast. He approaches a tree, walks quickly around it to judge the grade, bends over and saws through the trunk, lets the tree fall, marks the butt for grade with the edge of the chainsaw blade, and then moves on to find the next tree. The whole sequence takes thirty seconds.

Following Jim, collecting and dragging the trees to the side of the skid road that loops through this section of the plantation, is Dave Maskell. Dave is about six-three, probably two-hundred-twenty pounds. With his red wool shirt and gray wool pants, his thick dark hair, his pleasant but slightly fatigued face, he looks quite the lumberjack, even though he's just dragging Christmas trees. To keep up with Jim, he must sometimes break into a trot. Still, he always drags two trees, occasionally three, through the inch-deep snow: drags them, drops them at the edge of the skid road, then turns back down the row to find the next two or three. He rarely stops moving.

Meanwhile, the baling crew, Charlie Waterhouse and another Dave, Dave Allen, move along the road with the tractor and baler. They stop at each of Dave Maskell's piles. The baler, mounted on a long T-frame hitched to the tractor, is essentially a short, wide funnel sitting atop a gas engine. Charlie sticks the butt of a tree into the big end of the funnel; Dave hooks the end of the cable to it and pushes two levers. The cable pulls the tree through the funnel, compressing it, while a box of string zooms around the funnel's exit, spinning a tight spiral around the emerging tree. When the tree is wrapped, Dave disengages the two levers, unhooks the cable, ties off the loose ends of the string, and sets the tree beside the skid road to be picked up later. Then he reaches through to hook up the next one, which Charlie has meanwhile fetched and loaded up. They bale about a tree a minute.

They all move quickly in the brisk air. The temperature is in the twenties, cool with an overcast just thin enough to show a silver disc of sun. The men warm quickly. The sky slides overhead. The road from town remains empty. After his mumbled complaint about having his day messed up, Jim has worked steadily, concentrating on the task at hand. He looks roadward only once, at 10:15 A.M., when a pickup with a rack comes from the west. The truck passes by; someone else, heading elsewhere. At 10:30 Jim calls for a break.

The four men gather round the baler. Everyone opens thermos bottles and pours coffee.

"Who's got the donuts today?" says Jim. He looks at the others in mock expectation. No one has brought donuts.

"You don't usually give us a break," says Dave Allen, smiling. Dave, who looks to be in his mid-forties, has the wit and exuberance of one much younger. He smiles a lot, and seems untouched by the tedium of this job. His silver hair flows from beneath an old canvas baseball cap that is so worn that the fabric has split along the rim of the bill, exposing the cardboard structure beneath. The tatters along the crown don't quite obscure the logo: "Masterpiece Theatre." In the growing seasons Dave runs an organic farm. The rest of the year he helps out with logging and Christmas tree harvests and works at a cross-country ski center two towns away. Both he and Charlie work here partly because they like Jim, and partly because they want to see how he juggles his way through the Christmas tree season. They are both trying to establish tree farms of their own.

Jim says he's lucky to have people like these working for him. "You can't always find people who understand what I'm trying to do here," he says, swirling his coffee round his cup. "Who realize how important it is to me that this go well, and that the product is good – much less work this hard."

"It's not *so* bad," says Dave Maskell.

"Well it is highly physical work," says Jim. "It's about twice as physical as most people want to stand. You would agree with that?"

"Oh yeah," says Maskell.

"People these days have a distaste for hard physical work," says Jim. "I've had quite a few that just don't show up the next day."

"I'm surprised they show up at all," says Dave Allen. "They should know about you by now."

"Right!" says Jim, laughing. "They know better than to go working for Jimmy Moffatt, by God! Those shirkers, you couldn't drag them here with a car. By now, it's self-weeding! I'm a hard-ass-to-work-for bastard. I *want* to be, I'm *inclined* to be, I *intend* to be. So I am."

Dave Allen, grinning, is shaking his head No.

"It's not so bad," Dave Maskell says again.

"It's the lack of donuts that hurts," says Dave Allen.

Charlie agrees. They threaten a donut strike.

"All right – break is over!" says Jim, and they all laugh. As they put away their thermoses, Jim asks Charlie and Dave how many trees they've wrapped so far. Just under two hundred, they tell him. Probably another two dozen waiting at roadside.

"Call it two hundred, then," says Jim. "I'm going to take a chance here and assume Larry Demars is going to make it. You think you can get me another hundred fifty wrapped by lunch?"

"We can try."

Jim has cut another fifty trees when from the road comes a honk and he looks up to see Larry Demars's truck pull into the landing. The others stop dragging and baling, and begin loading the already baled trees onto the woods trailer to take to the landing.

Jim walks down and greets Demars. Demars is a squat, bearded man. He has a limp and, this morning, a sheepish grin as he apologizes for being late. He explains that he spent the morning trying to find a replacement for Al, the stand operator who'd been scared off by the Manhattan thug. He couldn't find anyone, so he finally returned to Al's house to try and convince him to go back. That took a while.

The rest of the news is good: The trees he bought the week before sold well, and he does want his three hundred more today. "Three hundred *at least,*" he says. "Maybe three-fifty. Whatever this truck will hold." The truck is a one-ton pickup fitted with tall siderails supporting a high rack that extends forward over the cab.

By now, the two Daves and Charlie have brought down the woods trailer full of Christmas trees. Long stacks of trees cut earlier in the week already lie at the side of the landing. Jim reviews the order with Demars, then calls for the crew to load.

"I need one hundred five- and six-foot balsam fir," he says. He and Charlie climb onto the truck. The two Daves fetch the first two trees from a stack next to the landing area. They hand the trees to Charlie and Jim, who load them. Every time one of the Daves hands Charlie or Jim a tree, he calls out the count, which the man taking the tree repeats. It is surprisingly easy otherwise to lose track. Demars, watching and listening, knows very well that this is the easiest opportunity to cheat a customer out of a few trees. Ugly disputes can evolve if the count strays. Jim tends to toss in an extra tree or two now and then to show he has no intention of shorting anyone.

After the fir come seventy-five short Scotch pine. Then fifty red pine, twenty-five blue spruce, and finally a hundred balsam fir under five feet. Short trees sell well in New York.

When the trees are loaded, it is just past noon. Jim turns towards the house and asks Demars to come in. They cross the Branch Road, bang their boots on Jim's porch to knock off the snow, and go into the kitchen. Joan comes in from the next room, says hello to Demars, asks him how he's doing. Demars, more subdued indoors, holds his hat and says, Fine, Miz Moffatt. Joan tells him it's good to see him again, then slips out to the next room.

Jim pulls a chair out for Demars at the kitchen table and asks if he'd like coffee. Demars says no thanks.

"Well I'll have some," says Jim. He pours himself a cup, black, and sits down at the kitchen table.

"Now," he says when Demars is seated. "You think you might want some more next week."

"I suppose I might," says Demars.

"Let me tell you what we can get you more of in a hurry," says Jim. "These are all smaller trees I'm talking about. Number ones and twos, fir. I've got fifty each of those on hand now. I can put you down for possibles on those."

"Possibles. Fine," says Demars.

"Some Scotch pine, fifty. And I can get twenty-five smaller blue spruce any time, premium. And of course I've got everything larger in most grades."

"I don't need too many over six feet. Not in New York."

"I know. But this is what I've got in the way of short trees at this point. I wonder what I can count on you for, next week." He pushes his hat back off his forehead. "What I should set aside for that last big weekend."

They both know what's being negotiated: how many trees Demars will have to put a deposit on for the next week, and how many Jim will definitely hold for him. The weekend that follows Demars's next pick-up will be the last big tree-selling weekend before Christmas. Neither man wants to be caught short or with extras when that weekend rolls around. It's the most volatile time of the season, when there might be a surplus of cheap trees around or a shortage of everything you need, and when, as Jim says, "Anything goes, there's no loyalty, and everybody knows it." Demars sits, thinking. He shifts his feet under the table. Jim waits.

Finally Demars says, "What you can count on and what I might want, might not be the same."

"Well," says Jim, "what do you want to count on being here?"

"Maybe a hundred. The mix you're talking about." He thinks again. "A hundred fifty."

"Say half fir, the rest pine and spruce?"

Demars thinks a minute. Nods. "Not more than twenty-five spruce, though."

"All right then," says Jim, smiling and tapping the table once with his fingertips. "We'll call it a hundred fifty. You can write me a check for that now, I'll hold them, and you tell me as soon as you can if you'll want more when you come get those."

"All right," says Demars.

"Good," says Jim. "Now we can eat."

He gets up, telling Demars to stay right there and make himself comfortable, and starts rummaging in the refrigerator. Finds a plate of leftover turkey. Joan comes in and heats some soup as Jim carves turkey for sandwiches and talks trees with Demars. The sandwiches and the soup are ready just as Charlie and the two Daves file in, bringing their own lunches in large paper bags. The table grows crowded; Dave Maskell carries a rocking chair in from the living room and sits down next to the woodstove. As everyone eats, Dave Allen tells the story of the time he and Charlie borrowed Jim's brand-new pickup. They switched all the stickers on the controls, so that for the next few days Jim would reach to pull on the lights and set the wipers flapping, or try to twist on the wipers and find himself spinning the cigarette lighter around and around.

"It was a long time before I lent that truck out again," says Jim.

When Dave Allen pulls an entire package of Oreos out of his bag, Jim, whose plate and bowl are empty, sits forward in his chair.

"Where you been hiding those?" he says. "You know better than to bring that food in front of me when I don't have any. I get meaner'n a bastard."

Dave Allen opens the package, pulls out two cookies, and holds them out to Jim. Everyone laughs and asks for theirs. There is crunching and happy mumbling as the group massacres the entire bag. Jim leans back in his chair and pulls a bottle of brandy and some shot glasses from a cabinet. He pours a shot for each person, caps the bottle and puts it back in the cabinet, then pushes the glasses across the table for his guests. All raise their glasses and drink. With the deal just done and the warmth of the woodstove – and the brandy, and the aroma of the soup, and the scent of Christmas trees on everyone – and outside a darkening sky shedding snow, it's beginning to feel a bit like Christmas.

The Christmas tree business exemplifies well the larger challenge that confronts Jim Moffatt. He goes by a simple name, "tree farmer." And there is a comforting, rhythmic regularity to much of his work. But this does not change the fact that he and his family run several different businesses, each subject to dozens of variables. At any moment, any of these businesses might be disrupted by changing international markets and trade regulations; rising local labor costs, property values, or tax rates; or major fluctuations in supply and demand resulting from bad weather, corporate decisions having nothing to do with trees, variations in forest health brought about by phenomena as nebulous as acid rain or as tangible as a tornado, or the changing condition of forests elsewhere in the nation or world. The world's wood markets have become so volatile that lumber prices can rise or fall by as much as 50 percent in a few weeks. Bizarre developments can cripple local markets. Sometimes a disruption comes out of left field – literally – as when the 1981 major league baseball strike snuffed northern Vermont's hardwood pulp market because the area's main buyer at that time sold most of its processed pulp to a nearby company that made most of *its* money manufacturing the little paper sleeves in which ballparks wrapped their hot dogs. No baseball, no hot dogs, no pulp market. Such disasters waylay the best of plans, yet the Moffatts must somehow try to account for every potentiality. They must make decisions daily about how to manage trees that they raise over decades to sell into markets that change by the week. If they guess wrong too often, they go under. In the end, says Jim, "There's nothing to do but do your homework and then go with your gut."

Jim's friend David Marvin, a forester and tree farmer from the nearby town of Johnson, twenty-five minutes west, feels the same way. Marvin too owns about 700 acres, both softwoods and hard, from which he pulls a mix of products similar to Jim's. Like Jim, he bought the land from his father years ago, and like Jim, he has managed it well. In 1984, he was recognized as the year's National Outstanding Tree Farmer by the American Forest Council.

Marvin does more than grow trees. He and his wife Lucy also own a maple-syrup brokerage business; a specialty store on Route 15 called the Butternut Country Store, which sells maple products and Vermont handcrafted goods; and a consulting forestry business. Marvin runs the forestry business from a first-floor office in a well-kept colonial on Johnson's main street, Route 15, directly across from the store. His desk sits in the middle of the room. Beneath a long bank of windows runs a low shelf piled with papers, technical manuals, and market reports.

"Trying to keep up," he says, gesturing at the shelf. "Twenty years ago, I knew exactly how to manage the woods. Now I don't know so much. There's so much conflict in the published information about small woodlot management.

"Take hardwoods, for instance. My feeling for a long time was that the best management was to harvest often, but always very gently. Now it appears that even working the stand lightly stresses the trees terribly. So I'm thinking maybe we should hit them harder but less often. I'm not sure. It seems there's always something happening. It's either thrips, or caterpillars, or a frost in the spring, or an ice storm or a wind storm. It's hard to know what to do.

"Take a maple stand you're wondering how to manage. The traditional line there is to make it eighty to ninety percent sugar maple. But a forest pest expert would say, 'We need to interrupt these stands with non-maples to avoid a monoculture. You should have twenty-five percent beech and ten percent yellow birch.' But *I* would argue that's no interruption at all! And in the meantime, you're spreading your stand out all over and increasing your work with tubes and taps. Then someone *else* might say that to have a really vigorous stand, you need to go in and thin frequently. Then someone else might point out – these are all experts, well-qualified people – that an outbreak of defoliating insects puts the greatest stress on a stand if the stand gets logged three years before or three years after the outbreak. Now, I can avoid harvesting a stand three years *after* an insect outbreak. But the three years before? I mean . . ." He throws up his hands. "These things all work at cross-purposes with each other. I have to take my own experience into real consideration. Otherwise I could never make a decision."

Occasionally when Marvin has a piece of timberland he wants to harvest and the stand requires a light touch, he hires Jim. In 1993, Jim and Steve cut a hundred thousand board feet of storm-damaged red oak on Marvin's property. Marvin made good money on the job, for it was fine oak, and oak was going for up to $700, or more, per thousand board feet. Jim and Steve, wanting a sure return on their time away from the farm, worked by the hour, charging a per-hour fee for their time and the skidder. Marvin and Jim share a lot of business. When one of them is short on one type of Christmas tree, he'll get some from the other, to be paid back in kind later. Jim also sells sugaring equipment for Marvin's equipment business, serves as a middleman in Marvin's bulk syrup purchases, and boils sap for him. They do most of their transactions verbally, leaving the reckoning until the dead days after Christmas, when they get together for an afternoon and figure out what they've done and who owes whom what.

Though small by big-city standards, the Marvins' businesses are critical to the town of Johnson, and are good examples of the sort of diversified, local businesses that are vital to a healthy forest-based economy. Altogether, several million dollars a year pass through Dave and Lucy Marvins' hands and into the hands of local foresters, loggers, landowners, craftspeople, and laborers. The Marvins directly employ twenty-three people full-time, and several others at least part-time through purchases of sap, maple sugar, cheese, and other food products, and Vermont crafts for the country store. (Just the little softwood boxes in which they package their maple candy products, for instance, provide two jobs and a demand for local wood products.)

The indirect effects of their businesses are also important. Dave's forestry consulting business, helps area landowners manage their lands well and directs wood toward local mills, which in turn sell much of their products to the many Vermont businesses, both large and small, that turn logs into products. Within a thirty-mile radius of Marvin's desk, nearly a dozen small mills saw wood into lumber and "bolt wood" (the small pieces from which wood-product manufacturers fashion a wide range of their goods); the Montgomery School House Company manufactures toys; the Newport Furniture Company fabricates furniture as well as those little jigs that help people cut bagels; Real Goods Toys constructs premium doll houses and a full line of other wooden toys; Tourin Musica builds stringed orchestral instruments; and Stowe Canoe and Snowshoe Company makes the obvious. Along with big manufacturers such as Ethan Allen Furniture, which employs thirteen hundred people in the state, these smaller firms – most of them employ a dozen or two dozen people – are critical not just to Vermont's rural economy, but to the existence and health of its forests, for they help create steady, local demand for wood of many kinds, from low-grade softwoods to the finest bird's-eye maple, thereby rewarding diverse, patient, and ecologically beneficial forest management. The diversity of the forest-based industry and the diversity and health of the forest are intricately related. In Marvin's words, "The more value-added business you have locally, the more refined the wood you can cut."

Jane Difley, executive director of the conservation group Vermont Natural Resources Council, agrees. Difley, who is a forester, was president of the Society of American Foresters during the early 1990s, and worked for many years before that in upstate New York and Washington, D.C., for the American Forest

Council. "If forest landowners know they have a strong local market at the end of the line," she says, "they're more likely to spend what's necessary to have good forest management. The other benefit you get is that the people who produce the final products become concerned about the resource and its management. If a big furniture company is located here in Vermont, as several are, and if they want to be here long-term, they've got to care about Vermont's forests being locally and carefully managed. And outfits like Dave and Lucy Marvin's play an important community role. The people who own the business are part of the communities, and contribute to the communities and work for them."

Difley, like many conservationists who understand the wood products industry, feels tremendous frustration that these industries don't get more recognition and support.

"I don't know if we have another industry in Vermont that's as invisible as the wood products industry. It tends to get ignored and undervalued. I wish the state would help them as much as they do high-tech business. You look at the Marvins's situation, for instance. The impact of their activities in those rural communities is extremely significant. *That's* what we need the state to be encouraging. To me, the Marvins have the perfect blend of the traditional forest-based economy, modern marketing, and high technology. They use computers and take advantage of the latest information and innovations, they put up a well-designed, attractive storefront – and they're using those things in service of a very traditional relationship to their land, their community, and to the small-town economy."

Payroll is the Marvins' biggest expense, followed closely by debt service and then property taxes on their 700 acres and the office and store in town. Theirs is a heavily leveraged, high-overhead operation. Dave Marvin doesn't speak of the overhead fondly. Yet he believes that the diversification of his family's operations, and the position further up the supply chain their leveraged processing and retail businesses occupy, provide a control he and Lucy wouldn't have if they just grew and cut trees. The diversification buffers them somewhat from market fluctuations, because they are getting a greater part of the total price of the products they make and sell, rather than just the money for the raw material. With any raw material, Marvin says, "the lower you get down the system, the less control you have. Kellogg has a much better chance of setting their profit margin than the Iowa corn farmer does. In fact, the farmer can't set his at all. All he can do is hedge the market. It's the same with trees and sap. As a grower, you can't set the prices. All you can do is try to read them and then maneuver to get the best margin you can. You have so little control. These smaller farmers . . . I don't really see how you can make a living just working a small wood-lot. I really don't see how it can be done."

The Moffatts are in precisely the position Marvin describes, particularly in their sugaring operation. Sugaring takes a tremendous amount of time. Boiling the syrup, which involves constantly feeding and tending the big evaporators that boil the clear sap down to thick sweet syrup, is actually the easy part. Each year Jim and Steve must change every single tap – they have ten thousand – and repair any sap lines knocked down by branches or animals, wind or snow. Yet the syrup market punishes or rewards the Moffatts for reasons unrelated to their own labors. The sap flows in quantities and qualities that are dependent not on how hard Jim and Steve work, but on some mysterious combination of weather factors. The market is further destabilized by the huge quantities of maple syrup being made in Canada. These uncertainties, and the general trend downward in sugaring profits, have driven the Moffatts to consider dropping the sugaring business altogether. Meanwhile, Jim and Steve are dismantling selected parts of the sugarbush so they can harvest some of the trees and release the rest for growing timber. Jim says that if he could do one thing differently over the last twenty years, he'd have phased back sooner on sugaring.

The Moffatts can exercise more control over their other crops. With Christmas trees, though the market is finicky, they at least create a finished, value-added product. In logs they have a bigger choice of markets to which they can sell, though that could change if markets get simpler rather than more diverse. The more furniture makers, small mills, craftsmen, and other value-added businesses around, the more customers the Moffatts have for the wood they grow; their future markets will be determined by how diverse a value-added industry Vermont can develop. If the local market for good wood were to expand, the Moffatts could, in a sense, add value to their trees simply by letting them grow until they are fully mature – though that requires patience.

Patience the Moffatts have. While they wait, they get by, as Marvin says, because they are smart and they hedge their bets well and they keep their debts low. Their strategy is to stay so lean that nothing can hit them. This frugality is not a concession; it's a weapon. Regarding the present Christmas tree glut, Jim says, "We know we're heading into a trough right now, one I think will last about five years. But I believe we're in the best of all positions to survive. We live frugally. We don't have a debt load. We think we can tough it out with the best of them."

Early in the morning on the last Saturday before Christmas, in a light, sporadic snow, Jim unloads trees alongside his retail stand at Bevin's Marine and Cycle Center in Montpelier. As he does every Saturday in Christmas tree season, he will spend this day selling trees to "city folks" – though Montpelier is not much of a city. At 8,500 people, it's the nation's smallest state capital, and the only one without a McDonald's, both facts that residents cite with pride. Nonetheless, Montpelier and its neighboring town of Barre are the main commercial centers in north central Vermont, and the Christmas season draws thousands of shoppers from the surrounding farm and bedroom communities. Bevin's location, on a secondary road running along the north bank of the Winooski River, is not ideal. Neither is the fact that its main business is boats – though the company does sell a few snowmobiles to help get through the winter.

To generate business during the Christmas season, the store's manager, Dee Codling, rents Jim most of the parking lot out front. She also lets a few local craftspeople, most of them retirees who needlepoint or knit or make small geegaws or trinkets, set up tables inside among the speedboats and snowmobiles. On this last pre-Christmas Saturday, Codling herself, an inveterate smoker in her forties with a whiskey voice and a wizened face, is dressed as Mrs. Santa. This fact has been advertised in the local papers along with Jim's trees and the crafts, but the main draw is clearly Jim. This is his second year at Bevin's. He sold 170 trees here last year and hopes to better that this year, despite heavy competition in the area. His location is inferior to those of a couple of tree stands closer to downtown and about half a dozen in parking lots and garden centers along the main suburban commercial strip, the Barre-Montpelier Road, which runs along the other side of the river. Yet he holds an advantage in price, for he has negotiated a low rent from Bevin's, and he is selling his own trees, saving the middleman's share.

The other advantage he holds is that Vermonters like to buy from farmers.

"There's no doubt about it," he says. "You *never* miss a chance to tell them that you're the one that grew these trees, and that you grew them on your own small farm. People want to buy from farmers. You'd be a fool not to let 'em know you are one."

Virtually every Vermont business that can do so – and a few that probably shouldn't – makes the Vermont home-grown claim a prominent part of its advertising. The power of this claim comes not just from Vermonters' recognition that buying locally helps their neighbors, but from Vermont's reputation as a place where small farmers still eke it out, where the environment is well cared for, where the forests are treated properly, and where there is a high standard of craftsmanship, whether in making furniture, cheese, train cars, ice cream, or bird houses. It's a reputation companies can readily export, as suggested by the many Vermont firms that identify themselves as such in their marketing efforts out-of-state. Nowhere is this more true than in Christmas trees. Many of Jim's out-of-state Christmas tree sellers, particularly those in Connecticut, New York City, and Long Island, advertise their trees as "Vermont Christmas Trees."

"You sell down toward New York," says Jim, unloading trees, "it pays to really yuck it up. I mean, you got to find your red plaid hat for that one. Lots of 'yeps' and 'nopes.'" He laughs. "It's just about impossible to overplay the part."

Here in Montpelier, Jim feels no such compulsion. He stays with his seasonal uniform of yellow nylon overalls, overboots, and a garish set of neon orange rubber gloves that he favors because they're hard to lose. His one concession to salesmanship is a Christmasy red-and-black plaid wool shirt and an unusually clean black baseball cap.

He's feeling good, smiling and joking as he unloads the trees, for he has had a good season, and his long stretch of fourteen-hour days is coming to an end. Most of his retailers have done well, and none are making the sorts of sounds dealers make when they're going to quit or go elsewhere next year. And Larry Demars seems to be turning into a steady and important customer. Yesterday he bought another three hundred and fifty trees; he bought eleven hundred altogether, at least three hundred more than anyone else. As of today Jim has sold almost six thousand trees, and expects to sell a couple more small loads over the coming week. It will be his best year since the late 1980s.

When a young, red-haired woman pulls up and gets out of her car, Jim gives her a smile and a hello as he pulls a tree off the truck. Without making a point of it, he makes himself available by setting the tree into the rack about ten feet from where she is standing. The young woman looks at the trees, still keeping her distance. She asks where they come from.

"We grow them on our farm in Craftsbury," Jim says, setting a tree aright. "My wife, my son, and I."

She asks if they're fresh. Jim tells her he cut them this week, "most of them in the last two days." She moves closer to the trees and rubs a bough between her finger and thumb, checking the softness. Jim stops rearranging trees and takes a single step closer to her. When she asks about price he tells her they go for $7, $12, $17, or $22, depending on whether they have a green, yellow, red, or blue tag. The red-tag trees are all fine trees, full and fairly even; the blue tags, the premiums, are as nice and symmetrical and full as a Christmas tree can be grown, and run six to nine feet. She takes a couple steps closer to the red-tag section. It's mostly balsam fir, with a few Scotch pine and blue spruce mixed in.

"Looking for any particular species?" Jim asks. "You like the balsam fir? Or are you a blue-spruce person?"

"My husband sometimes likes the spruce," she says. "I like the fir."

"It's a good native tree," says Jim. He pulls out two of the nicer red-tag trees, shakes them gently to help the branches fall a bit, and turns them slowly so she can see them.

"I was thinking something fuller," she says.

"That height about right?"

She nods. Jim puts the trees back and sorts through a few more, finds two that seem right. He holds the better one out, turns it slowly.

"That one's nice," she says. "That's a seventeen dollar one?"

"It is," says Jim. "Seventeen even."

"You take checks?"

"I do. You like this tree, you can make one out to Jimmy Moffatt. Two Fs, two Ts."

As she writes the check, Jim carries the tree to her car and ties it on top with some twine. She smiles at him as she hands him the check, and he nods and tips the bill of his cap, says come again next year, have a good Christmas. Then he returns to the racks of trees.

He repositions the trees he had moved aside a few moments earlier. He sweeps up the fallen needles, picks up a length of twine. Then he climbs back on the truck to unload the last of the trees. He takes each tree, bangs its butt on the ground to help the branches drop, and leans it carefully into the rack with the others. When he's done, the trees lean in neat rows against the racks. The pavement is swept clean. Only the truck, which is still parked at the edge of the lot next to the road with four trees in the bed and three leaning on the tailgate, looks like work in progress. He leaves those trees half-unloaded. It's his way of advertising: These trees are fresh, straight from the farm.

16

Off the Farm

Jim's brother Larry Moffatt lives in a farmhouse just outside of East Burke, Vermont, about thirty yards from the main road into town. Occasionally his house shakes as logging trucks gear down to deliver logs to the town's sawmill. Larry finds this sound soothing, for despite taking some hard knocks in the business, he still makes his living cutting wood. He has lived in the East Burke area since he was eighteen, when he moved the forty miles east from Craftsbury to attend college in nearby Lyndonville, but he has lived in this house for only three or four years. The house is a rental. He moved here when the logging empire he built during the 1970s and 1980s collapsed, slowly but violently, taking with it his house, his land, his logging equipment, and his second marriage. Now he lives alone. His two sons are grown.

When the house shakes, the glass-fronted china closet in the kitchen shakes too. On the top shelf are a photo of Larry's mother, Ila, and one of his father, Bob, looking old, gaunt, and stern, as Scottish as they come, solemnly playing the violin. Another shelf holds the trophies Larry and his dad won at the Craftsbury and other fiddling contests in the 1970s. "I played a lot for a long time," he sighs, but says that when his business fell apart, he quit. It was, he says, part of the sadness he felt in those years. "I could hear stuff on the radio, but it was mechanical. My soul didn't hear it."

He makes coffee. He spent the morning checking on a logging job in Danville, twenty miles south, where his son Mark and three other men are cutting maple on a 219-acre woodlot Larry recently

235

bought. Larry wears jeans and a long-sleeved navy blue tee-shirt; his hair is mussed from his cap. Perhaps because he spent much of the fall recovering from prostate surgery, he looks weary. His features are more blurred than Jim's, an impression heightened by Larry's wandering eye. He speaks slowly, deliberately. When he wants to drive a point home, he looks straight and steadily at the listener. When he tries to recall something or discusses something difficult, such as the dissolution of his business, he leans forward, elbows on the table, and screens his face with his hands, touching his fingertips gently to his forehead and closing his eyes as he thinks and talks. He rubs his face. But he lights up when he talks about working the woods.

"You always have certain expectations from what you've heard, or you remember what the wood's like over there. And then you get there, and if you walk into something that's different from what you expect, either good or bad, you just get so excited. To see and figure, How did it happen to be like *this?* Why are these trees here and those over there? What went on on this land over the last fifty or a hundred years? The challenge is to see what happened before, because that's how you learn about what to do now. It's an exciting thing.

"It's a great way of life, woods work. You can always make ten dollars, as long as you're willing to spend eleven. Anybody that wants to log can barely make it till they die, is the way I see it. But it's a great way of life."

The first forest Larry cut belonged to his father. It was a spruce stand in what the family called the Dustin swamp, the last big piece Bob Moffatt added to his property. Larry was fresh out of high school and looking for something to do. There was talk of him going to college. "I was, 'Well, maybe,'" says Larry. "The one thing I *knew* was, I didn't want to stay on the farm. I got to the point I hated it. I *hated* cow manure. I hated having phosphate in my face

236

out in the field. I hated the old-fashioned ways my father did things. And we disagreed about everything. Oh I was a real – I knew everything! I remember gathering sap. Dad had a Ford tractor, but he was a horse man, so he wanted to stay with the team at sap time. He and the hired man would work the horses, collecting buckets, and he'd give me these other routes to work with the tractor. He gave me the farthest ones, because I could make more ground. We'd leave the sugarhouse at the same time. I'd go alone, and they'd go the two of them, and I had it in my mind that I could beat them every time. That was my mission. And I often would. I was always like that. That was what I was about."

When Larry graduated from high school in May of 1952 and his father needed someone to cut the spruce in the Dustin swamp, he asked Larry if he and a friend, Pat Davis, would be interested in doing it that summer. It would be a contract deal, so much per cord. Larry jumped at the chance.

He and Pat set to work, armed with a chainsaw that Larry says "was one of the first chainsaws around." When they were lucky it ran one day a week. They felled most of the trees with a two-handled crosscut saw. They also peeled the bark, using an ax, because the mill paid a premium for peeled logs.

"And man, you want an unpleasant job," says Larry, "it's peeling wood in a swamp in June with the black flies. As soon as it was light enough to see we'd start to work, because it was cooler then. We'd work till about noon, when it would start to get hot, and then we'd climb under a tree and sleep. At four when it started to cool again we'd do some more and then we'd go home. And you know what? Then we'd go out every night! Oh, every *morning* I'd tell myself, If I live through this, I'm never doing it again. Then I'd do the same thing. Because I was eighteen.

"I liked it. We were making good money. It was our own project. Some people told us that swamp couldn't be cut. But we did it." The two young loggers cut the trees to fall toward each other, overlapping in herringbone fashion and propped up by branches so they could dry. They left the tops on the trees so that the needles would pull moisture from the trunks.

In August they began chopping the branches off and sawing the trunks into logs. Larry was standing on one of the trees limbing it with an ax when the blade glanced off and cut a deep gash in the side of his foot. For the next few weeks he was laid up, moping around the house, "feeling sorry for myself and wanting to get back to work because I knew I had this commitment. Pat was still working on it, by himself. He was not pleased."

All along, Larry had been paying his mother for room and board, at his own insistence. He continued to pay her while he moped and grumbled. As September approached, his mother asked him if he ever thought about going to school. He mumbled that he didn't know. She asked him what he would think about going to Lyndonville Teacher's College. He said he didn't know, he hadn't thought about it. She asked him if he'd be willing to interview there if she went with him, because she knew the president of the school and thought that Larry would like her. Larry said, Sure, why not?

"So we went. I liked the school fine. On the way home from the interview, my mom says, 'You know, you'll need money for tuition.' I didn't have a dime. I hadn't saved any – I was eighteen! She says, 'I saved all the board and room money you gave me. It's yours for school if you want it.' I never forgot that. I was like, Wow – how could anybody ever have a nicer mom than that?"

Larry enrolled in the teaching degree program, went for three semesters, then dropped out to do an Army tour in Germany because he wasn't sure he wanted to be a teacher. After his stint he re-enrolled. He married during his last year at school, then graduated and took a teaching job.

Soon after he started teaching, he and his wife bought a house with fifty acres. The fifty acres were full of cedar. Larry began cutting on weekends and school breaks. He started cutting cedar elsewhere too, paying for stumpage on other properties. After a year he realized he liked cutting wood more than teaching, so he quit teaching and started cutting full-time.

One of the companies he supplied was a Massachusetts-based outfit called Concord Woodworking Company. The company was impressed that Larry could find good cedar when other operators couldn't, so as the company began thinking about expanding, it offered him a position doing an inventory of the northern white cedar in the area. He took the job. Over the next three years he learned two important lessons about himself: that he knew more about the area's forest than he realized; and that he was not a company man.

Not that he didn't get along well. The company liked him enough to name him woods manager after a year, and he enjoyed the work and the steady money. But after three years the company wanted to move him to the mill.

"They wanted to put all the production in the mill on incentive plans," he says, "but they had no knowledge of how long it took to do what. I guess because I was the only one up here with a college education, they thought I could do just about anything. So they decided I'd be the guy to figure out how long everything should take.

"I'll tell you what, I hated that. I didn't want to be inside the mill. I didn't want all the guys hating me because I was standing there with a clipboard and a pencil calculating what they were doing all day. I mean, I could do the job. I saw that you could cut out steps and make things more efficient. That part was challenging. But just working in the mill was distasteful to me. I'm a woodcutter. I want to be outside."

So he quit. "Never looked back. I was never hung up on security. Because I always knew that if things went bad today, I could work twice as hard tomorrow and in about three days from now I could catch back up to where I should have been."

And so it went for years. Larry worked hard, fell behind many times, and many times caught up. He slowly added equipment –

a skidder, a truck, another skidder – and hired logging crews. His first marriage came apart. He remarried.

In the 1980s, a chip contract with a new wood-burning power plant in Burlington, Vermont, gave Larry the chance to expand more rapidly than ever. The MacNeal plant, which was owned by a consortium of interests including the Burlington Electric Department, was the first major biomass plant to open in Vermont. At 50 megawatts, this was a big facility, and it represented a major commitment toward establishing what was being touted in northern New England as a renewable energy wonder. Biomass plants are fed not by imported fossil fuels or radioactive substances, but by wood chips, which could be made from the low-grade wood so plentiful in New England's second- and third-growth forests. As alternatives to power plants fueled by expensive oil, gas, and coal, biomass power plants represented a big step toward energy independence for New England, which had no indigenous fossil fuels but plenty of trees. In addition, by providing a market for chips, such plants would make it economical for landowners to weed out their low-grade trees, thus improving their woodlots. It seemed to be a happy marriage of good economic and good environmental policy.

Larry was more than happy to do his part. A contract with MacNeal, he figured, would give him steady, guaranteed demand, providing a baseline from which to expand his business. More important, this security would let him expand his land holdings, which already included several hundred acres. He could buy stands of skinny low-grade hardwoods, thin them to release better trees to grow into high-quality logs in years to come, and chip what he cut for short-term profit. In 1983, Larry signed a contract with MacNeal. On the strength of that contract, he borrowed hundreds of thousands of dollars to buy land and whole-tree harvesting equipment. He soon had thirty people working in the woods, building roads and cutting and chipping and trucking wood.

For a while, business went as planned. Then several circumstances began to work against Larry and against biomass plants in general. The biggest factor was that energy became cheap again. By the mid-1980s, the weakening of OPEC (the Organization of Petroleum Exporting Countries – the mideast oil cartel) reduced oil prices to levels not seen since the early 1970s; gas and coal prices followed. In addition, huge hydropower plants in Quebec went on-line and added to New England's power supply. These changes completely transformed the role of the MacNeal plant within the New England power pool, the regional association of producers that shares power sources and profits. The MacNeal plant had been conceived as one of the pool's "base load" plants – one that would run all the time. Any energy it produced beyond local needs would be sold to the pool for use elsewhere. But because other energy sources had turned out to be cheaper, MacNeal ended up operating as an intermittent "cycling" plant, running only when regional energy demands were high.

The plant's owners, as members of the pool, shared in some of the pool's overall profit whether the plant ran or sat idle. Larry Moffatt, however, did not; he sold chips only when the plant was running. Larry says this possibility simply wasn't made clear to him when he signed the contract with MacNeal. MacNeal, of course, had no incentive to emphasize such a contingency, which seemed remote when Larry signed. MacNeal's cycling status caught everyone by surprise. The banks that financed the expansion of Larry's operation and those of other wood-chip suppliers, for instance, had done so assuming that MacNeal's power would be cheap enough relative to other fuels to keep the plant running steadily. Yet the economics shifted. In addition, objections in Burlington to smelly piles of decomposing wood chips prevented MacNeal from laying up too much chip inventory, so the company couldn't buy steadily even if it wanted to.

Larry was caught. Though his agreement provided no compensation when the plant wasn't buying, it required him to stand ready to supply the chips. He couldn't sell his land and equipment without risking being sued by MacNeal for breach of contract. He found himself with several thousand acres and hundreds of thousands of dollars worth of equipment – "a shearer, a feller-buncher, about five skidders, two chippers, and about six big trucks – expensive things with a lot of payments, a lot of headaches." To buy all this, he had borrowed over a million dollars; his monthly loan payments ran to the tens of thousands. Yet he was seeing little or no income from selling chips.

To make up the difference he took the only course available: He cut for every market he could find, and when that didn't produce enough cash, he cut harder. Unfortunately, other markets were sporadic and provided only marginal returns. The most prominent other market at the time was for pulp, but the mills didn't want the relatively dirty, low-grade material that Larry's on-site chippers produced.

None of this pleased Larry. When he had bought the several thousand acres, which were in patches around northeastern Vermont, he had planned to hold them indefinitely. He built good road systems that would allow him to thin the stands periodically to produce ever-improving stands of maturing hardwoods. But with his profit margins so narrow and his debts so huge, he found he had to cut his lands harder and harder. "I had no choice," he says. "I did a few things then I'd rather not do. Made some clearcuts that were quite visible. In the overall scope of things they weren't big cuts – probably nothing more than ten or fifteen acres. But people didn't like to see them."

In the end, even cutting hard wasn't enough, for without steady demand from MacNeal, neither chips nor pulp produced enough profit to carry Larry's overhead. He fell further behind. Eventually he had to start selling land, often for less money than he needed to recover what he'd put into roads and other improvements. Each sale gave him cash to hold off his creditors for a few months. In the meantime, he prayed that wood prices would go up before he went under.

It was, he says, a painful time.

"I spent nineteen eighty-six, eighty-seven, and eighty-eight dying financially," he says. By the end of the decade he had sold virtually all of his land, his house, and most of his equipment. The game was up. He sued MacNeal to get free of his contract, and negotiated with his creditors. The lawsuit released him from his contract with MacNeal in 1992. That year he also came to terms with his creditors and was able to start slowly rebuilding.

But his life was not the same. For beginners, his second marriage was over.

"This forestry and logging business is not great for families," he says. "The idea for both Linda and myself that we weren't able to pay our bills was devastating. You've got the sheriff at the door every three days – it's a nightmare. I really don't want to talk about it a lot, because I'd rather not relive those years. But when you've been through what we went through, it's devastating. I guess the easy way to say it would be that if you're familiar with the term post-traumatic stress syndrome, that's what occurred. Some people who haven't experienced it don't understand. But it's *a very real thing.*"

For Jim, watching his brother go through this was a painful experience. Larry, by his own admission, was an angry, embittered man during this period. He railed at MacNeal. He railed at the government agencies that regulated – or, as he saw it, failed to regulate – the energy industry. He railed at anyone who seemed to stand in the way of his efforts to recover from the blow he'd received. At times, some of those close to him worried that his bitterness would overwhelm him.

It did not, he says, though it left him permanently affected.

"I'll never be like I used to be," he says. "I don't trust anybody. I used to trust everyone, until they did me in. Now I don't trust

243

anyone. I don't laugh and cry like I used to. I don't get mad and I don't get excited. I'm just . . . I'm missing a whole bunch of emotions. They're not there."

These days Larry makes his living growing and selling Christmas trees and cutting wood. With land prices what they are, he can't afford to hold land after he works it. Instead he buys, cuts, and then sells. He says he manages these lands the same way he would manage them were he keeping them for the long term. And that may be. But this pattern of buy, cut, and sell worries a lot of people concerned about the Northern Forest. They call it "cut-and-run." In its worst manifestations, this pattern leads to cuts driven solely by monetary concerns – by people, as fishing guide Dale Wheaton puts it, "whose only concern is how much money they can make off the land between the time they buy it and the time they sell it."

Essex County, where Larry lives, has been the site of several such operations over the last few years. Most of the heavy cuts in Vermont have been done on lands owned by logging contractors who buy them specifically for that purpose. Parcels several hundred acres in size have been clearcut and high-graded – stripped of all but the worst and least valuable trees – and then sold, often for development. One operator in particular, a logger named Larry Brown, has bought thousands of acres (some of them available because of the Diamond land sale of 1988) and clearcut or high-graded them. Some of the cuts are at high altitude, or on steep, thin, or otherwise vulnerable soils, which will take decades to recover. The wood and related jobs these woodlots might have produced over the next few decades are meanwhile lost, as are the wildlife habitat, erosion control, and soil temperature stabilization they provided. Some of the clearcuts are near Route 2, the main east-west highway through northeastern Vermont. They shock and sadden many who pass by, including quite a few loggers, such as Larry Moffatt, who have sometimes clearcut smaller tracts themselves.

Larry says what bothers him about the big Essex County cuts is their scale. "I'm not crazy about the idea of cutting more than twenty or twenty-five acres. Something that big, you should at least strip-cut, leave some rows of trees to shade the ground and keep it cool so you can control the temperature and get some seed for regeneration. Those cuts along Route 2 I can't abide. I never drive that road that I don't think about it. I'm quite disturbed about it, to tell you the truth."

Many environmental critics and foresters also see clearcutting as a problem not because of its inherent characteristics, but because it is so often done on a large scale or in the wrong places. As for regulating such forest practices, however, Larry Moffatt parts company with the environmentalists.

"We've got some problems in the Northern Forest," he says. "But one Larry Brown in all of Vermont is not a real big problem. It just happens to be on Route 2 where everyone can see it. What I worry is that this will become an excuse to lay a bunch of plans and regulations on everybody like me and Jim. What they do to stop Larry Brown could kill us. I'm worried you won't be able to cut a tree unless you have a permit from Forester Smith, and that Forester Smith's ideas and mine won't coincide, so I won't be able to cut enough trees on the land I own even to pay taxes. That's what I worry about."

Larry doesn't oppose all regulation. He approves, for instance, of Vermont's existing forestry regulations – which are, he admits, fairly lenient, addressing mainly cuts along streamsides and at high elevations. Yet he worries about any regulations that might come out of regional or federal agencies. His concerns about regulation are like environmentalists' concerns about clearcutting – he worries not so much about the phenomenon itself as about its scale. Like many northern New Englanders, Larry's aversion to regulations increases with the distance from which the regulations emanate. Regulations coming out of his own town are one thing; out of the state capital, Montpelier, another; out of Washington, D.C. another altogether. His big nightmare is "some blanket program that starts out in Washington as a good thing and turns into a monster when it gets here to come down on us." He doesn't have to look further than the Occupational Safety and Health Administration regulations to which he is subjected each day to find examples of monsters.

These fears are consistent with the traditional Yankee insistence on independent home rule – an insistence that sometimes strikes outsiders as provincial. The attitude is based on a notion, inlaid in age-old tradition and powerfully evocative because it speaks to true and basic needs, that only those who live and work in rural communities can comprehend the values and dynamics that make those communities viable, valuable places to live. Rural people recognize that large legislatures dominated by urban and suburban representatives and influenced by business lobbyists do not understand the concerns and strengths of rural communities.

Larry Moffatt says, "It's sad to think that a lot of folks think we're not capable of taking care of this place. We are. We're doing a damn good job of it, and a much better job than we were thirty or forty years ago. We may sometimes cut certain blocks or do certain things for reasons others might not understand. But for us to destroy our own forest would be like a farmer murdering his cows. We're going to be more protective of these forests than anyone. We earn our livings from them.

"People from away, they think Vermont is something special, something unique. And it is. But they don't have any inkling of the pressures on guys like Jim and me who keep it so wonderful so they can enjoy it. We're the ones that are responsible for looking after this place. We're the ones here every day doing it. So we'd better have some input into what comes down on us. Because human beings can get seriously hurt by this kind of stuff. We're on the bottom of this kind of thing, Jim and I. You're kind of floating around in a zone where you don't have to get hurt. But we can."

Larry has begun to play the violin again. He has also been reading about violins, for he has long had an interest not just in playing, but in the instruments themselves. "I used to collect them," he says. "I still have a few that are a couple hundred years old." Right now he's reading a biography of Stradivarius, which sits on the kitchen table next to a cassette player holding a Waylon Jennings tape. Sipping his coffee, Larry talks about the great violin maker, explaining that Stradivarius's genius was in carrying forward the work of his own teacher, Nicholas Amati of Cremona, and refining the crude instruments of the time into the more elegant, sweeter-toned "Cremona" design that has been the standard ever since. He talks of Giuseppe Guarneri, the other great violin maker of Stradivarius's time. He observes that their violins were made with the same species of wood, spruce or pine tops and maple backs, ribs, and sides, that are readily available here in Vermont, and that over the years there have been a couple of companies in Vermont that made violins. But those instruments, like those made everywhere else over the past three centuries, have failed to match the perfection of the violins made in Italy in the 1600s. No one knows why the seventeenth-century Italian violins have proven so superior to almost all subsequent violins, even those of identical dimensions and species of wood. Their superiority seems to be the result of some almost metaphysical combination of construction technique, the composition, application, and drying of the oil varnish, and, most mysterious of all, the seasoning of the wood, which in a violin must be dried and warmed just enough that the oil varnish will readily penetrate its pores, preserving their elasticity, but not so much that the wood becomes brittle. All of this fascinates Larry immensely; he talks about violins with the same delighted animation with which he speaks about forestry.

He tells of going recently to a chamber music concert given by a Japanese couple and their six-year-old daughter. After the show,

Larry went up to examine the violins. They offered to let him play one. He could see the instrument was valuable and asked how much it was worth. Slightly embarrassed, emphasizing that it was on loan to them, the couple confessed that the violin was worth two and a quarter million dollars. Larry hesitated, afraid that he would drop it. But when they held it out he took it and played for a moment anyway. The violin, made four thousand miles away and three hundred years before, had a spruce top, maple sides. It sounded superb. He did not drop it.

17

Passing the Saw

A few days before Christmas, unloading one of his last truck-loads of trees, Jim Moffatt slipped on ice. He had just grabbed a tree to pull it off the back of the truck, and his right hand stayed tangled among the branches as his feet went out from under him, putting his shoulder through a terrific twisting strain. In the following weeks the shoulder hurt badly enough to keep him awake almost every night. He got through that last week before Christmas season only by eating up to two dozen painkillers a day – Tylenol while he worked, Percoset at bedtime in a vain effort to sleep. In the idle weeks after the holiday, when he had the time to actually feel the pain, he wondered how he had worked during that holiday week at all.

His doctors told him to rest, but rest didn't help. In January he went back to work cutting maple, using his right arm sparingly and leaving the heavier work to Steve. When mud season came, liquifying the roads and ending the winter logging season, he tried rest again. The pain persisted. "He's so stubborn," said Joan. "I don't think he complains enough. I don't think the doctors even realize how much he hurts."

When almost four months passed with little improvement, Jim's doctors decided to operate. They knocked him out and snaked a pencil-thin cable holding a fiberoptic lens and tiny instruments into his shoulder. They found that he'd torn a good part of his bicep loose from its origin within his shoulder. They cleaned up the accumulated scar tissue to create a site for reattachment, sewed the muscle back to its fibrous anchor, and, emphatically this time, told Jim to rest, his arm in a sling, so the shoulder could finally heal.

Now, in May, a month later, Jim no longer wears the sling. But neither does he wield a chainsaw or lift anything heavy. He will try to gradually resume full activity in July. Soon after that he will know what his arm can do. In the meantime, while Steve shears Christmas trees, bringing the next season's trees into rough shape before they put on their summer's growth, Jim cruises his woods and decides what he'll cut this year. Maple prices are still good; he thinks he may cut more of his sugarbush in the fall. He sets up some contract logging jobs off-property for late summer. In the afternoons, he catches up with friends and reads: two Civil War histories, of which he has read many before; *Merry Men,* one of Carolyn Chute's novels about backwoods Mainers; Michener's *Texas.* He enjoys long lunches with Joan, drinking coffee on the patio overlooking the Wild Branch. They talk about the farm.

The injury has sharpened their awareness – made plain as day – that Jim has only a limited number of years in which he'll be able to work as hard as he has for most of his life. He has maintained for three decades a physical pace that taxes men in their twenties, and the work is taking its toll. Steve, meanwhile, is showing an increased interest in the farm; it's becoming less a matter of *If* and rather one of *When* he will take over. Jim's injury has reminded the family how much remains to be settled about how to handle the transition.

"The basic problem," says Jim, "is to work out a plan where Joan and I can be financially secure in retirement, and yet not burden Steve with something he can't handle and maintain a relatively pleasant lifestyle. I went through some awfully difficult years here, and I wouldn't wish that on him. I wouldn't encourage the situation if I thought that might happen. Though it really takes two people to do this. A single person just can't carry it off. You'd have to hire a secretary, a treasurer, a receptionist, an errand person, a salesperson, all rolled into one. It's cheaper to marry one!" he says with a laugh. "So quite frankly, that is a problem.

"The other big problem is that we'll need some money to retire on. I don't know how to handle that."

Joan says, "These things worry us a lot more now that Jim's health seems a little iffy. We worry how to make it work so we can all live decently. It *is* hard to see how Steve could handle this. All this land. I mean, you have to practically give up your life to run this place. It's a hard thing. You want to leave him in a situation where you *know* he'll get along. Then you realize at some point he'll have to take his chances like the rest of us."

The financial challenges of passing the farm to Steve are daunting. If Steve buys the farm at market value, he will never be able to pay the loan each month with what the farm yields. And if his parents simply give it to him or leave it to him in their wills, the gift or estate taxes might force Steve to sell a good part of it. In Vermont, federal and state estate taxes typically take at least half of everything beyond the first $600,000 in value of an estate. Jim figures that the farm's seven hundred acres – the hundred thousand Christmas trees, the four to five hundred acres of pulpwood, sawlogs, and sugar trees, the house and barn, all the equipment – are worth anywhere from one to one-and-a-half million, depending on how development potential and timber values are assessed. So if Steve inherited the property under ordinary circumstances, he might face a tax bill – due within one year – of up to half a million dollars. To raise that kind of money he'd have to cut hard on large tracts of the land, sell the most developable parcels, or do both.

Selling some choice parts of the farm, of course, is an option now as well. Much of the Moffatts' land, both in the bottomlands along the Branch and in the gently sloping sugarbush overlooking the valleys south and east, is a developer's dream. They could sell one hundred, two hundred, three hundred acres, and in one fell swoop provide money to fund Jim and Joan's retirement, reduce

Steve's eventual estate and property tax loads, and shrink the farm to a size manageable by one person. They had the chance to do just that in 1992, when the Lamoille County Solid Waste District, looking for a new landfill site, cast its eye on the Moffatts' property. "What it came down to," says Jim, "was that we could sell about half the farm for about five hundred thousand dollars. *That* would have made a nice nest egg." But the Moffatts told the District they weren't interested. "It just didn't seem the right thing to do to the land, or to our neighbors," says Jim. "Didn't seem like a decent way to treat the land that we thought a lot of."

Nevertheless, the Solid Waste District's offer opened Jim's and Joan's eyes about the value of their land and raised the issue of what to do with the farm. To explore their options, Jim and Joan went in March to see a Burlington lawyer who specializes in estate planning. The lawyers laid out possibilities so numerous and complex as to be numbing. "It's a very complicated area," says Jim, "full of arcane strategies. But if you don't do something, things just happen to you. We've seen enough people around here get caught by surprise in that situation. It's a terrible thing to see."

They could, the lawyer explained, sell conservation easements on all or some of the land, taking cash from the state or a private conservation organization in exchange for surrendering any option to develop the land later. Selling these easements might bring the Moffatts anywhere from $100,000 to $250,000 in cash and allow them to continue to work the land; and since any land on which they sold easements could never legally be developed, it would also lower the estate's taxable value, and thus reduce or eliminate Steve's eventual estate tax liability. The down side, of course, is that the easements would lower the sales value of the property, reducing the Moffatts' ability to raise cash if needed. And even if they did want to pursue this route, private and public dollars for buying easements are extremely limited.

Another option would be to donate part of their land to a charity and, through a complicated arrangement called a charitable remainder trust, have the charity provide Jim and Joan with income from the land for the rest of their lives. This too would reduce the taxable estate, although the Moffatts would lose the use of the donated land immediately and completely. The lawyer described other trust options as well, both revocable and irrevocable, and various measures it would take a week to understand.

Jim and Joan pondered their choices for a month. Then they decided that for the time being they would change the farm's ownership to a joint-tenancy-in-common – a trust arrangement that essentially gives Steve the benefit of the $600,000 estate-tax exemption twice, for a combined total of $1,200,000 of exemptions. As long as the farm's value doesn't increase too far beyond that amount, this should keep Steve's tax liability fairly low. The joint-tenancy arrangement also leaves the family's options open as they explore other possibilities, and as Steve further tests the idea of staying on.

"It may be that Steve falls into other opportunities in the meantime," says Jim. "I think he deserves a few years to . . . well, if something exciting happens along, no reason he shouldn't do it. I've never pushed him to come back, because those things have to happen by themselves. Now he says it was always in the back of his mind that he would come back here by the time he was thirty. I don't know why it couldn't be a good life."

"I'm sure he'll do fine," says Joan. "Especially when we're out of the picture!" She laughs. "He gets impatient sometimes. It's not easy, working every day with someone you're that close to emotionally. I remember going through the same thing when Jim's dad was here, before Jim had the farm. There are tensions. But we never worry that Steve won't do the right thing. He's a hard worker, and he seems to know what it's going to take. He wanted to go do something just last night, some program about wolves he was interested in, but with all the shearing to be done, he said, Well, I guess I can't make it."

253

In the old red barn next to the house is a high-powered motor-cycle Steve rides into Burlington now and then. He skis, and spent a week last winter schussing the Italian Alps. He has fished in Alaska and hunted in Wyoming. He already enjoys pleasures his dad never knew. Jim and Joan don't begrudge these pleasures, perhaps because Steve seems to accept, if not yet completely embrace, the amount of work required to run the farm. Steve has been working since he was a kid, maybe not as hard and as young as Jim did, but much harder than most kids his age.

In his high-school class of twenty-one people, says Steve, "there were basically two groups." He's alone in the sugarbush, taking a break from cutting to sharpen his saw. On the ground next to him are his hardhat, a can of extra fuel, a plastic bottle of chainsaw oil, an insulated nylon lunch bag Joan gave him last Christmas, and some of the sap lines and taps he has pulled from the trees he's thinning. As he sharpens the chain, the quick strokes of the file sing into the clear day. "There was the redneck group, that hunted and went on to auto mechanics school. And there was the college-bound group. Except for me. I was going to go on to college, but I did the work of the rednecks. I used to shovel the horse barns when we had those, and I always split a lot of wood. One summer I sold wood with one of our neighbors. Then starting high school I was responsible for all the sugar wood – that's about a hundred cords right there – and all the house wood: cutting it, splitting it, stacking it. I always had chores.

"But I don't remember thinking it was that much. Most of my friends worked too. One grew up on a dairy farm. Now he *worked.* The only time work ever interfered with anything was when I started playing soccer, in the seventh grade. Soccer practice started a few weeks before school did in the fall, and it was at three-thirty every afternoon. I remember my dad saying, 'What's this world coming to when a boy's summer chores end early so he can go to soccer practice.'"

That Steve returned to the farm after college is notable, for most kids who go from rural areas to college don't come back. The blue-collar kids, the sons of loggers and farmers, stay to work the woods and repair cars; the college kids move on. College degrees may not be quite the meal ticket they used to be, but they still provide economic prospects rosier than those typically found in towns like Craftsbury, even if your parents own a farm. You don't do it for the money. Steve doesn't talk at length about his own motivations. He tends to steer the subject toward trees. ("Did my dad tell you what we're doing with this sugarbush?") He seems hesitant to discuss any plans he might have for the farm. It's hard to tell if this is because his plans are vague, or because he feels it's presumptuous or inappropriate to discuss what he'll do with his inheritance, or if he's simply reserving judgment until he sees how his and his parents' lives work out. Most likely it's a combination. But when you press him, it's clear that he has given the question some thought.

"I could make as much or as little of this as I want. It would be *fairly* easy for one person to get by and make a living here. But if I'm married with a family it will have to be a lot more involved, unless she has a tremendous job. Either way, I'll probably have to do less than we're doing now."

He sets the file on a stump and stands up, puts on his hard hat. "One thing I *wouldn't* do," he says, flipping his face guard down and picking up his saw. "I wouldn't put in ten thousand taps."

In the cooling hours of a fall day, with the leaves a riot of flame in the slanting afternoon light, Jim drives his truck up the road that rises into the hills north of the house. He passes among fields, drives past a nicely restored house that a young couple from out of state bought a few years ago and fixed up; they're in the garden, and Jim waves. The road tops a rise and crosses a broad hilltop, an old sheep farm where the freshly mown hay fields lie bright green against the autumn trees. Then he descends through the brilliant hardwoods. The leaf shade slides over the hood and up the windshield. He crosses the small brook that forms the border of his back five hundred. After another quarter mile he pulls left into a landing, which is full of spruce and fir and a few hardwoods that Steve and Jim dragged out of the sugarbush in the spring, marginal logs they had cut to release some of the maples.

Steve is already there, standing next to the logs. Jim gets out. Father and son nod at each other, smile. Jim pulls a tissue out of his pants pocket, rips two pieces off, rolls them into balls, and stuffs them in his ears. "My official, high-grade, OSHA-approved hearing protection," he says. Then he pulls his saw from the pickup. His arm, after being sore off and on all summer, has been feeling better the last few weeks, and though he hurt his back in July, it too has firmed up in recent weeks, and today, he says, he's feeling pretty good. Before him the trees lie more or less parallel on the ground, their tops still attached, still holding the long-dead leaves. Jim starts his saw and quickly buzzes off the top of the tree nearest the truck, then starts cutting the trunk into 4-foot sections. Steve, holding a pair of 8-inch hooks, sinks a hook into either end of the first 4-foot section and loads the log onto the pickup.

They work this way a couple of hours, Jim cutting, Steve loading the cut sections of log either onto the truck or into neat stacks. The logs in the truck will go to the sugarhouse to fire the furnace that will boil next spring's sap. The stacked logs will be picked up for pulp.

It's a good day to work outside. Jim and Steve quibble occasionally, arguing good-naturedly over whether a log is fit for pulp or just a sugar log, or where to put the next stack. Finally Jim is done cutting. There's still about half a pickup load of logs to sort. The logs lie just to the wrong side of a hummock of dirt and a stack of pulpwood with about a dozen logs in it.

"Too bad that pulp is in your way," says Jim. "Otherwise you could just back right up to those logs. Save a few steps."

Steve, hooks in hands, looks at the pulp. He looks at the truck, at his dad, and back at the pile.

"I'm not moving it," he says.

"Well it'd save you a lot of work. Would have, anyway."

Instead, Steve gets in his pickup. He pulls forward a few feet, driving the front wheels onto a short stack of logs, then revs the engine and tries to back over the hummock of earth. Doesn't make it. Jim shakes his head. Steve pulls forward and puts the truck in four-wheel drive. He tries it again, revving the engine higher. This time he backs over not just the mound but the two junk logs behind it and puts the truck's bumper three feet from the logs he wants to load. Then he steps out of the truck, trying not to smile too much, gets his hooks, and starts loading again.

Jim, smiling too, gives his son the universal look of admiration reserved for sons who know just how smart-ass to be.

"I'm glad we agree on everything," he says.

18

The End of the Season

With the Christmas tree stand closed for the night, Jim Moffatt gets in his pickup. He crosses the Winooski River and then turns left onto Route 2 for the drive north to Craftsbury. He feels good. With only a week left before Christmas, he is starting to feel the relief and space the end of the tree season will bring.

"If we get a nice cold snap after Christmas," he says, "I'll go ice-fishing as soon as I can. I don't ice-fish that much. But I like to do it. I've been up to Memphramagog a few times. Had one *excellent* catch there. I generally go for the perch, because I like to eat them.

"But not too many people like to do it with me. I'm pretty numb when it comes to cold. Steve says, 'When you go with Dad, take your own car, because he just won't leave.' It's true. I had one friend I went with last winter, we took my truck and he 'bout froze to death. My wife was talking to him on the phone a few days later and I hollered to her to ask him when he wanted to go fishing again. He told her, 'You tell him to go to hell! Tell him go straight to hell!'"

He laughs, remembering it.

Outside the cab a half-moon lights the snow beside the road. We follow the Winooski to East Montpelier, then take the split in the road that sends Route 14 north toward Hardwick and then Craftsbury. Jim leans against the door as he drives, his hat pushed back off his forehead; he's wearing his yellow overalls and the red and black plaid wool shirt. The headlights push through the night. We somehow get talking about bears, and Jim tells a story about a family that bought a cabin on a ten-acre piece of land along the Moffatt Road, just across from his sugarbush.

"We called them the Bear Hunters," he says. "They were from New Jersey or Connecticut or some place, and they'd come up weekends. They were scared right to death of bears. Scared out of their wits." He starts laughing. "They were nice enough people. You'd go up there, they'd give you a beer, they'd be glad to see you. But I mean they were *really* scared of bears. They'd say, 'Heard something moving around the camp last night. Must have been a bear.' And they'd say, 'You haven't seen any bears lately?' And we'd say, 'Oh I don't know,' make something up. Because at first we thought they were asking because they liked to hunt bears or maybe wanted to see some, and we didn't want to be too discouraging. Finally we realized they were always asking because they were so scared. Well. After we caught on, we really performed for them. All of a sudden we were seeing a *lot* of bears.

"I guess it worked too well. They got stuck once in their car, stuck in the woods down below the camp, and they wouldn't leave the car. They were scared the bears would get them."

Told that the story is reminiscent of the short stories of E. Annie Proulx, the Vermont writer who illumes with such accurate irony the divisions between newcomers and natives of northern New England, Jim says, "Proulx is one of the writers that gets it. She's one of the ones that actually understands rural people. Carolyn Chute is another. She is very good at expressing how things really are for those people. When we read the work of someone that tries to talk about this local native population, we pay a lot of attention to how sensitive they are in picking up on those people's feelings and thoughts, what makes them tick. And those writers do. But there are some that don't. Some that really miss it."

Asked to name one, Jim says, "This Budbill, for instance." David Budbill is a poet and playwright who lives in the Craftsbury area and writes about Vermonters, some of them his and Jim's mutual neighbors. "He has a great following," says Jim, "among the,

259

shall we say . . . well, you tell me what's a fair word for a person, someone who's arrived on the scene in the last ten or fifteen years. What do you call those people? Something that's respectful. We'll call them newcomers. To the newcomers, Budbill's descriptions of these lower classes are amusing. But he can't quite hide his snobbishness. He looks down on them. Which to me is *terribly* unfair. Some of the people he's talking about in his book, thinly disguised – we can pick them out easily, the ones we know – are people that worked for my father, and I remember them very well. A lot of them, their children worked for me. And he just doesn't do justice to these people. There are other writers like that too. They just don't put their fingers on the local culture. They miss it. They think they've got it, but they miss it. They don't show respect for who these people are. To me, every one has to respect everyone else in order to understand each other. That's why you can't say certain things. If you feel them, fine. But you can't say them."

The same failure to grant respect, he says, haunts relations among people in town. He feels if people grant each other respect even through their mutual lack of understanding, they can bridge the gap. But he recognizes that people repeatedly fail to do so. He's seen how incidents such as zoning or Sterling College's barn or big school budgets can start to split towns apart. Everyone gets along fairly well in Craftsbury right now, yet Jim worries that the next great conflict, which could be anything from a teacher's pay increase to a big subdivision to efforts to resolve the regional problems affecting the Northern Forest, could inflame prejudices, befog understanding, and break down the town's sense of community. He worries that newcomers fail to try to really understand the community they move into. He worries equally that those who have lived here all their lives will refuse to adjust to the fact that things change. He sees too many locals closing ranks, blaming the struggles they face on the regulations that newcomers and other meddlers, as they see it, impose on their freedom. Jim has one friend, a nearby farmer who does some logging – one of the most intelligent men Jim knows and one of the most personally generous, "a good friend, a good neighbor" – who also, Jim says, "has the most unusual interpretations of ideas I've ever known.

"There's no right, left or middle – he's *all* the way to the right. He thinks that all state and federal regulations should be removed, that people should be able to cross streams with skidders and trucks, it don't make no goddam difference. He and I disagree on that. The truth is that in the Northeast, here in Vermont in particular, we have more freedom than the Declaration of Independence ever gave us. More security, a better life than our forefathers ever intended or hoped for us to have. When you corner these people and ask them, 'What regulation is it that's offending you so? What's taking away your freedom?' They can rarely nail it down, with the exception of those that want to sell real estate, and they of course complain for very selfish reasons, because regulations like Act 250 make it harder to sell land for an exorbitant profit. But Act 250 doesn't stop you from doing anything you shouldn't be doing anyway. We've all seen the development that's overcommercialized. It's the end of the community. The benefits of these regulations far outweigh their drawbacks."

He slows for the one light in Hardwick and turns left. We follow the road along the Lamoille River, past The Checkered Flag restaurant and bar, past Windshield World and the Rite Way Sport Shop and a gas station selling "Xmas trees," closed for the night. We pass the House of Pizza and Poulin Lumber and the Arctic Cat snow machine shop and then the Grand Union and the Chittenden Bank and the Ford dealer and Hardwick Motors. A smeary rain begins to fall. We drive into the darkness outside of town and turn right, north again, and drive alongside the thin long white expanse of Hardwick Lake. Then we enter the beautiful, narrow valley that leads up to Craftsbury. The road runs serpentine between steep hills that are dark against the sky. Jim yawns. The rain turns to snow. For several minutes he seems either lost in thought or sleepy. Then he suddenly picks up right where he had left off.

"Of course, my wife and I are not in the majority. We're frugal people. We don't require much to be comfortable or happy. We believe happiness is in yourself. We believe that you as an individual have the power to structure a pleasant life for yourself. We lose patience with people who feel, 'Oh, if they just had fewer regulations, everything would be fine.' Or if they had more money, or a better car. And so forth and so on. How good do they expect things to get?"

Too often, he says, these people blame their problems on newcomers. "They just see no good at all from the newer people, and all bad." He purses his mouth into a tight bitter scowl. "Oh," he says, in a mingy voice, "they'll want to be on the school board, and they'll want this, and they'll want that." He laughs at his caricature. "You hear that all the time.

"But then, the newer people often don't understand those of us who've been here for a while either. They try, sometimes. They'd often *like* to. Some of my better friends, actually, are people who moved here from elsewhere. They're people I respect a lot. They seem to have a lot of respect for me and for what I do. But understanding?" He shakes his head. "Don't think so. Some of that may be my own fault. I don't like the rest of the world talking on about how everything is such a hardship, so I don't talk myself much about how hard I work and how tough it is and all that. I've shown you because you wanted to know. But I try to downplay that.

"But these people, some of them, they tend to romanticize my life a bit. They see me and my wife as having, A, a perfect marriage, and B, a perfect business, and they think that anyone would be highly privileged to step into my shoes. And 60 percent of that is correct. I am privileged. But they have *no idea* of what it entails from the time we get up in the morning to the time we go to bed at night. It's awfully hard to explain to someone, you come home at night, say someone's there dropped by for a visit, you come in the door and your clothes are all torn up, there's blood dripping off your nose where you banged it on something, you're all tired, you're hunched over because your back hurts . . ." He laughs and shakes his head. "It's awfully hard to explain to someone that Yes, this is the way it is, but Gee I had a good time today."

We're past the turnoff to Craftsbury Common now, and up to the Branch Road. Jim turns left and swings past the Collinsville store, past Moffatt Road. The headlights catch the trees at the ends of his Christmas tree rows, which line the pale hill in dark rows. He slows and pulls into the driveway. Joan has hung a wreath on the door since we left this morning, and the kitchen lights are on. Jim gets out of the truck, leaving the keys inside, grabs his jacket and shuts the door.

"Well," he says, "Saturday night. See what's happening here. Have a couple highballs. Eat dinner. Read my book. Go to sleep. We'll hit it again tomorrow."

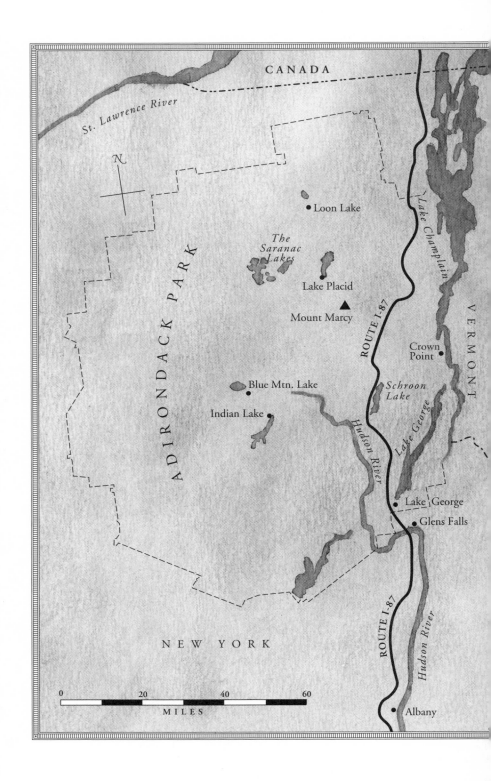

CANADA

St. Lawrence River

N

Loon Lake

The Saranac Lakes

Lake Placid

Mount Marcy

Lake Champlain

ROUTE I-87

VERMONT

Crown Point

Schroon Lake

ADIRONDACK PARK

Blue Mtn. Lake

Indian Lake

Hudson River

Lake George

Lake George

Glens Falls

NEW YORK

ROUTE I-87

Hudson River

0   20   40   60

MILES

Albany

Part Four

NEW YORK AND BEYOND

—

"...We who live here know that things
cannot stay the same.
But we do not want to be treated as side-issues.
We do not just live and work
and play in the forest — we are part of it.
We are ingrained in it, and it is ingrained in us."
JOHN HARRIGAN,
North country newspaper editor

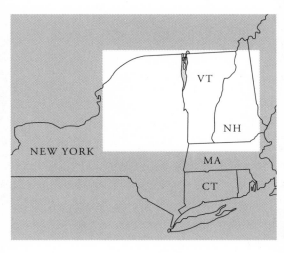

AREA OF DETAIL

19

The Adirondacks

Carol La Grasse spoke at only one of the twenty public hearings held by the Northern Forest Lands Council in the spring of 1994 – the last hearing, held in Glens Falls, New York – and she spoke for only three minutes. But no one left a stronger impression.

Before her, fifteen other speakers had been introduced by moderator Ben Coe and taken their allotted three minutes to give, in generally civil terms, their opinions on the thirty-three draft recommendations the Council had produced. For the Council's seventeen members – environmental, government, forest landowner, and local representatives from each of the four Northern Forest states – this listening session marked the last stretch of a long road. For three years they had struggled to meet their directive from Congress and the governors of Maine, New Hampshire, Vermont, and New York to find ways to "reinforce the traditional patterns of land ownership and use" in the Northern Forest. They had strained their eyes reading studies, taxed their hearing and patience at dozens of public meetings, driven all over the region to meet with citizens and interest groups, and finally, late in 1993, met in a marathon weekend session to try to reach consensus on a set of conservation proposals. It was these draft recommendations to which citizens were responding at the listening sessions, and the citizens were telling the Council that it had done a decent job. For a consensus document necessarily compromised by divergent interests, most agreed that this was a good start. The document was entitled "Finding Common Ground."

267

There were criticisms, of course. Some environmentalists complained that the recommendations regarding acquisition of public land, protection of biodiversity, and regulation of forest practices were too accommodating of industry, and urged the Council members to strengthen these proposals in their final report, which was to be published in the fall of 1994. Critics from the forest industry complained that the sections regarding government land purchases were too vague and asked the Council to specify how much land should be protected. And virtually everyone testified that the measures aimed at strengthening local forest economies were unimaginative and did not go far enough to aid either wood-based manufacturing businesses or recreation and tourism. Overall, however, the reaction could be accurately called positive-with-reservations. Most speakers thanked the members for making a good effort on a tough job.

But as soon as Carol La Grasse walked to the front of the room, in long, forceful strides, a copy of the Council's report under her arm and a heavy purse over her shoulder, it was clear that she had not come to pat the Council on the back.

"This is a disgrace," she said when she got to the microphone. She turned to the four Council members from New York. "Mrs. Sweet, Mr. Stegemann, even Mr. Bendick, you ought to have some sensitivity about it, and even *you*, Mr. Woodworth, to serve on an entity that's caused untold heartache and fear among our ordinary people." Bob Bendick, who as head of New York's Department of Environmental Conservation (DEC) had weathered harangues from angry citizens before, stayed motionless in his chair, his hand cupping his chin. Neil Woodworth, the conservation director of the Adirondack Mountain Club, did the same, his hands on his thighs. But Barbara Sweet, a councilwoman from the small Adirondack town of Newcomb who considered herself sympathetic to local interests, and Bob Stegemann, who as an official at International Paper was used to being criticized by environmentalists but not by angry defenders of property rights, looked alarmed at being singled out by name. Sweet and Stegemann pulled back their heads, eyes wide, as La Grasse walked straight toward them. Leaving the microphone behind, she advanced until only a row of empty chairs separated her from the Council members.

"This," hissed La Grasse, waving her copy of the draft recommendations at Bendick and Woodworth, "is the ultimate corruption of government! When government is a tool for terrorization of decent people who want to live out their lives. Whose attachment to the land puts you goddamn environmentalists to shame. Hunters who care about the food on their table. Ordinary working people who really *do* work in the woods, who really *earn* their living with the forest that you pretend to want to preserve so that you can put your *goddamn animals back!* Instead of presenting your forest lands study, instead of presenting your 'Common Ground' report, you should have apologized to the people of the Adirondacks. You should be ashamed to use the power of government, *to use your mills,"* she said, turning to wave the report at Stegemann, who pulled his head back even further, eyes opening wider, as La Grasse advanced until she stood only three feet in front of him, "to cause terror . . ."

La Grasse dropped the book. It fell over the row of seats and landed at Stegemann's feet. "Give it to me!" said La Grasse. She shoved the intervening chair aside and reached down for the book. Stegemann pulled his legs out of the way. She picked the book up and retreated a step, tucking the book under her arm.

"I need the damn thing for the next time I have to write a goddamn report about you people who destroy ordinary people's happiness," said La Grasse. She began pacing back and forth across the front of the room. "I hope the gates of *hell* open for you," she said, her voice rising as she looked at the Council members, "for the terror of just *one person* who called me yesterday morning and said, 'Carol La Grasse, have you read the *Post-Star,* do you see what they have in store for us now?'" She turned on her heel. "The ordinary people, the slobs, who live in hovels, who have to worry about the goddamned junk car ordinance, who have to worry about the feds with a greenline system for twenty-six million acres. Sure, you cleaned it up. You don't have the entire program that you had originally. No, you wouldn't *dare* submit the whole zoning plan that you had in the original bill . . . And you know it, Bendick," she said, turning and approaching him, "you lied to the farmers . . ."

"Miss," said Ben Coe, who had risen and gone to the microphone, "this is not common courtesy."

"No," she said, turning. Coe, a white-haired septuagenarian with a kindly and slightly nervous manner, tensed at her approach. His left hand gripped the microphone stand. "No, it isn't common courtesy," La Grasse went on, "when people call me in the morning in terror. Now I'll tell you – let me have your mike and I'll tell you what I have to say." She tried to grab the microphone stand from Coe, but Coe stood firm. She pulled at it once, hesitated, then turned back to the Council.

"This report," she said, "should have had two sentences: I'm sorry. I apologize to the people of the Adirondacks and the people of the North Country for writing the Northern Forest Lands Study and for conceiving all of these studies and causing all of the terrors of government."

"Time is up," said Mr. Coe.

She turned back to him. "I'm sure my time *is* up," she said, pointing a finger at him. "Yours is up, too, but in a different way."

And at this, as about a dozen of the room's eighty people applauded, she turned and walked along the front wall of the room to one of the exits, slammed through the door, and was gone.

La Grasse was not the only Adirondack resident to speak against the Council's proposals that evening. While most of the forty-one speakers gave favorable or mixed reviews, almost a dozen Adirondack residents, including a pinstriped real estate lawyer as well as Dale and Jeris French, the blue-collar leaders of a property rights organization called the Adirondack Solidarity Alliance, blasted the Council. They said the Council was out of touch with local needs, that the recommendations went against the New England and New York tradition of home rule, that the Council was bent on turning the entire region into a playground for the rich – even asserting that the whole process was a conspiracy in which environmental groups, the forest industry, and government had joined to take land away from Adirondackers.

The harshest critics repeatedly sounded the property rights activists' central complaint, which was that the Council's efforts trod upon property owners' constitutional right to use land as they saw fit. Some speakers wandered through rambling monologues. Others read concise statements. Some appealed to intellect, some to emotion. One woman, Bonnie Howard (who was married to David Howard, a leader of the national property rights organization Alliance for America) hushed the room when, voice shaking, she said that for the last three years while the Council worked, she had felt like a hostage. "We sit here tonight," she said, "and talk about a place for the trees to grow and the deer to run and the tourists to visit. I want to know, What about a place for me and my children to live?" She then asked everyone there "to look at my face . . . and . . . promise me that if I lose my home to acquisition, you will exchange your home for mine. My home is on the line. I'm asking each of you to put your home on the line also."

Howard expressed very well the fear, rational or not (the Council had publicly rejected eminent domain as a means of acquiring land), that lay behind many Adirondackers' objections to the Council. But no one expressed better than La Grasse the anger that rose from that fear, and which had crippled the Council's work in the Adirondacks. At the Council's first meeting in New York state, held in the town of Ray Brook in August 1991, angry Adirondackers had vandalized two Council staff members' cars and threatened the Council members in such hateful language that several were visibly shaken.

271

The ferocity of these objections, which was unique to New York, rose from what was both the upstate region's greatest environmental triumph and its biggest political albatross: the history of the Adirondack Park, six million acres that is roughly half public land (2.6 million acres) and half private, in the Adirondack Mountains in the northeastern part of the state. As the largest park of any sort in the continental United States, and as an unusual mix of private and public land, the park was considered a regional treasure to outdoor enthusiasts and an exemplary experiment in land-use policy. As the country's most prominent attempt at "greenlining" (the practice of defining a region in which special land-use regulations and programs apply, an approach used widely in Europe but in only a few places in the United States), the Adirondack Park had been mentioned early on as a possible model for the larger 26-million acre Northern Forest study area.

That was before the vehement objections of New York property-rights advocates had helped convince the Council that greenlining would not be politically feasible. This resistance had spun off from the complex, one-hundred-year history of the park and, more immediately, from the bitter, twenty-year dispute between Adirondack residents and the state's Adirondack Park Agency, which not only managed the park's public land, but also regulated land use and development on the private land encompassed by the park's boundaries. Many locals believed these regulations had become not just inconvenient but actually hostile to their existence. "They're trying to shut down America," said Adirondack Solidarity Alliance leader Dale French in reference to the Park Agency, the Northern Forest Council, and similar regulatory efforts. "The people who feed, clothe, and house America are being shut down, their resources being locked up. It's an all-out offensive against the people of this country."

272

The tension between use and preservation in the Adirondacks dates back more than a century, to the post-Civil War years in which heavy logging – part of the wave of aggressive timber harvesting that was occurring throughout the Northeast at the time – led to the first appeals to protect the Adirondack mountains by means of some form of park. Steep, cold, inhospitable to farmers, the Adirondacks had been largely passed over by settlers in America's westward expansion. Yet loggers did not ignore the great stands of mature forests so close to the nation's manufacturing and trade centers. They cut with abandon through the middle of the nineteenth century, shipping the logs down the Hudson River in great rafts and, later, on the rail lines that penetrated the Adirondacks in the 1860s and 1870s.

The same rail lines carried visitors who saw other values in the Adirondacks' fine scenery, wildlife, and clean air. Throughout the 1860s and 1870s, these tourists and adventurers were drawn to the Adirondacks by the magazine and book writings of people whose very names – Verplanck Colvin, Ned Buntline, Seneca Ray Stoddard, and the Reverend William H.H. "Adirondack" Murray – hinted at the exotic nature of the mountains that the railroads put within a day's journey from New York City, Philadelphia, and Boston. Even more so than the White Mountains of New Hampshire, which are actually higher, the Adirondacks, with their steepness, their tight valleys and alpine heights, their heavy spruce and fir forests and beautiful lakes, gave – and still give – a sense of a remote, enclosed mountain kingdom. They offer the visitor tundra-topped alpine summits, deep undisturbed forests, some of the finest trout streams in the nation, scores of beautiful alpine lakes, and a mix of form and light so distinctive it inspired an entire approach to landscape painting, the Hudson River school. In the increasingly industrialized society of post-Civil War America, this newly discovered region seemed a treasure of wildness and respite – and a dreadful thing to waste. "Within an easy day's ride of our great city," noted the *New York Times* in what would prove a seminal 1864 editorial, ". . . is a tract of country fitted to make a Central Park for the world. Let [us] . . . seize upon the choicest of the Adirondack Mountains, before they are despoiled of their forests, and make of them grand parks . . . where at little cost [citizens] can . . . seek the country in pursuit of health or pleasure. In spite of all the din and dust of furnaces and foundries, the Adirondacks, thus husbanded, will furnish abundant seclusion for all time to come; and will admirably realize the true union which should always exist between utility and enjoyment."

The aggressive harvesting in the region also troubled the emerging fraternity of scientific foresters, who worried that clearcutting and resulting erosion would destroy the forest's future productivity, and hydrologists, who even then recognized the link between healthy forests and water quality and supply. Hunting and fishing groups were also alarmed at the forest's destruction. The most influential constituency to grow concerned, however, was the New York City business community, which worried that irresponsible timber harvesting was damaging the Hudson River watershed. The Hudson not only supplied New York City with water, but was a vital shipping lane for both raw materials going south to the city and finished products going north to the Erie Canal and then westward.

The heavy logging of the Adirondacks during the middle of the 1800s thus galvanized a powerful alliance, uniting preservationists like John Muir who wanted to preserve the area's beauty, conservationists like Gifford Pinchot who wanted to see more farsighted management of the forests, urban recreationists who wanted to enjoy the area's beauty and solitude, and the business community, which wanted to protect the water supply. By the 1870s, these diverse interests had joined in making a single, powerful argument: that maintaining the environmental health of the Adirondacks was a statewide and (through its importance to New York City, the nation's center of commerce) national priority that justified special protections. At the time, the Adirondacks were thinly settled, so local sentiment was only a minor factor – the main struggle was between large timber interests, which wanted to freely exploit the area, and the conservation coalition, which believed that larger regional interests superseded the timber companies' right to do as they pleased.

The conservationists' argument proved persuasive. In 1885, the New York legislature, concluding that "private ownership means – sooner or later – forest destruction," created a "Forest Preserve" in the Adirondacks consisting of all the state-owned land in the region. This amounted to just over one hundred thousand acres initially, most of it in patches of land that timber companies had clearcut or high-graded and then surrendered to the state for taxes. The legislation called for this "Preserve," which was actually a collection of lots scattered throughout the Adirondacks, to be kept "forever wild." In 1892, the legislature responded to calls for a more formalized park in the region by declaring that both the original Forest Preserve and all other state holdings in the thirteen Adirondack counties would thereafter "constitute the Adirondack Park." More important, the 1892 legislation designated a "Blue Line" that surrounded not only all the existing Forest Preserve land in the Adirondacks (five hundred thousand acres by then), but also the private land between and around those holdings – 2.8 million acres in all – and called for the state forestry commission to concentrate on acquiring more land within that area to expand the park. The act didn't otherwise mention private lands. But the drawing of the Blue Line forever defined the Adirondack Park as a distinct region, both public and private, in which the rest of the state had a declared interest. In 1894, the state's voters made the "forever wild" clause part of the state constitution, thereby making the Forest Preserve the only constitutionally protected land in the nation. Over the next century, the Adirondack Park grew steadily as the state repeatedly expanded the encompassing Blue Line and bought more land for the Forest Preserve.

The park's reputation also grew, as the beauty of its untamed terrain, its proximity to New York City, and its designation as a region "forever wild" gave it an almost mythical identity. The emergence of the Adirondack "camp" in the late 1800s was a crucial part of this reputation. These "camps" included everything from simple cabins that middle-class people could rent to palatial estates owned by America's elite. No matter what their level of luxury, most were marked by rustic furnishings meant to exemplify the ruggedness of the terrain and the adventurous, healthy lifestyle their occupants came there to pursue. "Sanitoriums" and other early versions of today's spas and health resorts were also popular. The very airs and waters of the region were said to be physically and morally restorative, whether you were a backcountry explorer or an ailing consumptive. Hunters, hikers, poets, presidents, the sick and the healthy – all found something here they could find almost nowhere else in the East.

By the late nineteenth century the Adirondacks had become one of America's premier vacation destinations. This popularity waned somewhat as beach vacations came into vogue in the early part of the twentieth century, and as cars and then planes made exotic destinations more accessible. But the Adirondack Park has remained a favorite retreat for visitors seeking "rusticity" and solitude, whether it be a day playing golf at a lakeside resort, a weekend fishing out of a creekside cabin, or a week hiking in the backcountry.

By the 1960s, eight million people a year were coming – as many tourists as were visiting Yellowstone. This popularity, along with America's post-war prosperity, led to resort and second-home development that slowly but surely began to change the character of the park's private lands. The area around Lake George, for instance – the first large Adirondack lake reached by anyone traveling from New York City – became ringed by inns, resorts, and vacation homes. Ticky-tacky development so overwhelmed the town of Lake George (on the lake's southern tip) that locals began calling it "The Coney Island of the Adirondacks."

The development boom led environmentalists and many residents to call for growth control. In 1968, Governor Nelson Rockefeller, a Republican, responded by appointing the "Temporary Study Commission on the Future of the Adirondacks" to recommend ways to limit, or at least plan for, growth. In assembling this fifteen-member commission Rockefeller made a blunder – an easy one to make at the time – that helped set the tenor of Adirondack politics for the next twenty-five years. He assembled a group dominated by patricians, community-minded business people, and environmentalists, and almost completely lacking any "grassroots" Adirondackers. Few objected at the time, for as Frank Graham, Jr., points out in his excellent *The Adirondack Park: A Political History,* no one expected this committee's work to go any further than many bureaucratic commission endeavors do – which is nowhere at all. And in view of the park's history, the commission's composition probably seemed appropriate, since this represented the same alliance of interests that had created the park in the first place.

The vital difference between the alliance that created the park and the Rockefeller Commission, of course, was that the latter group was explicitly setting out to create policy for private lands.

The members of the Rockefeller Commission took their assignment quite seriously. They worked for almost two years, employing a staff of twelve and generating over fifty reports. Their final recommendations, "The Future of the Adirondack Park," published in 1970, proposed the creation of an Adirondack Park Agency (APA) that would not only administer and expand the existing Forest Preserve land, but would "have planning and land-use control powers over private land in the Park." The Commission also drafted an omnibus bill for the 1971 New York legislature that would create the agency, give it zoning and land-use control powers, expand the Blue Line, and buy more land for the Forest Preserve.

With the governor solidly behind this bill and enjoying majorities in both houses – and with the environmental movement springing into life across the nation following the first Earth Day in 1970 – the APA Act passed pretty much as the Rockefeller Commission had written it. The bill's backers made two significant concessions to their opponents. The first, made in response to objections from municipal officials in the Adirondacks that their authority was being lost to the APA, was the establishment of a Local Government Review Board, or LGRB – a board of locally elected officials that would "monitor and advise" the APA; the LGRB was given no authority, however. The other concession was the softening of proposed regulations on shoreline development in exchange for severely limiting development in much of the private backcountry land.

In essence, the law made the APA a regional zoning board, with authority over a wide variety of projects. The agency's zoning and land-use map designated six different types of private lands, each restricted to a certain density and type of development. Growth would be largely uncontrolled in the existing "hamlets," or population centers. But in the other five land types, a developer or landowner couldn't build without a permit showing that the proposed project met the relevant zoning and environmental considerations. Progressively decreasing densities and increasing environmental restrictions would be applied to development in "moderate intensity use areas," "low intensity use areas," "rural use areas," and "resource management areas." There were also "industrial use areas" where existing industries could remain and new ones locate. The scope and density of any given project might be further restricted by the APA's assessment of the project site's particular environmental characteristics, such as soil type, slope, and groundwater conditions.

From the beginning, the agency's authority over building activities chafed many locals. Many residents believed that the political process that produced both the agency and the land-use plan had ignored them. They based this belief partly on the failure of the state to grant the Local Government Review Board any authority. They also considered themselves under-represented on the APA review board, despite the requirement that five of the eleven board members be Adirondackers, because those five were gubernatorial appointees rather than elected local officials. Further, critics said that the APA had inadequately consulted town officials before creating the zoning map, giving them only a few weeks to study the complex map and apply for changes in their towns before the map became law.

The situation didn't improve when people started trying to get permits. The guidelines were complicated, the process unwieldy. A developer or landowner with a project in mind first underwent long meetings with the APA in which the staff told the applicant which density and environmental guidelines to address – but not, with any specificity, how to address them. The vagueness of this pre-plan "consultation" made it easy for an optimistic developer (is there any other kind?) to develop a certain notion of what would be allowed; put together a full proposal based on that notion (often a long and expensive process, since proposals had to be very comprehensive and sometimes required accompanying environmental impact statements from geological, biological, and/or hydrological consultants); and then either have the proposal rejected outright or accepted with so many conditions that it became a different project altogether. This happened not only on big projects but on small ones as well. Even well-intentioned developers found the length and unpredictability of the process frustrating.

The APA staff, of course, were in a tough position. They were charged with enforcing a growth control act explicitly but imperfectly designed to control both large-scale development and the piecemeal "thousand cuts" variety that can slowly ruin a landscape and damage the environment. This latter concern led the staff to review many minor projects with maddening rigor. In one early decision they denied a man named Joe Hickey permission to live in an apartment that he had built, with local zoning approval, over his store. The staff ruled that the second use would give Hickey two principal buildings at a site where APA regulations allowed only one. The agency eventually backed off and let Hickey move in. But such cases – Hickey's became a *cause celebre* in the growing property-rights movement – seemed to confirm accusations that the APA was run by anti-growth greenies who had no clue of what it's like to try to make a living.

In another early case, the agency tangled with a developer named Anthony D'Elia. D'Elia had been working since 1970 to revive a defunct resort on Loon Lake, one of a chain of lakes in the northern Adirondacks. The resort had once been a prime destination for America's ruling class (in fact it had been known as the "Summer White House" around the turn of the twentieth century, and had hosted five presidents), but had since fallen into disrepair. D'Elia wanted to refurbish the existing inn and golf and tennis facilities and add an 840-unit subdivision. He had re-opened the golf course, demolished some of the old buildings, begun restoring others, and built three miles of new roads when the APA came into being in 1973. Because D'Elia hadn't "substantially undertaken" the permitting and construction of the subdivision, the APA put that part of the project through a full permit review – including, as the result of a snafu, two separate environmental impact statements, each costing tens of thousands of dollars. Altogether, D'Elia's permit review took almost three years and cost him legal and consultants' fees that he claimed ran to a quarter of a million dollars. When he eventually won approval in 1975, the permit was, he wrote later, "laced with so many unexpected, illegal and unreasonable conditions of approval that I knew I was finished." He abandoned the project.

This treatment so infuriated D'Elia that he devoted the rest of his life to trying to do to the agency what he felt the agency had done to him: put it out of business. In 1980 D'Elia would publish the book that became the history as well as the manifesto of the Adirondack property-rights movement, *The Adirondack Rebellion,* but in 1975 he was still a key part of that history. In February of

1975, he joined forces with Ruth Newberry, a wealthy woman who objected to the APA's regulation of her several thousand acres of Adirondack land, and a few others to form the League for Adirondack Citizens' Rights – the first of many citizens groups devoted to abolishing the APA. Some of the league's members soon helped found the Adirondack Minutemen, who had the same abolitionist cause, and the *Adirondack Defender,* a quarterly newspaper that recorded the APA's "crimes." Throughout 1975 and 1976 these organizations held rallies and meetings at which they railed at the park agency and called for its dissolution.

In 1976, D'Elia gained membership in and then the chairmanship of the Local Government Review Board (LGRB), the advisory body set up by the APA Act and charged with "monitoring and advising" the agency. The LGRB was composed of local citizens appointed by the supervisors (who themselves were elected) of the thirteen Adirondack counties. For the first three years of its existence, the LGRB, headed by an Indian Lake resident named Dick Purdue, had served as a shaky but vital bridge between local interests and the APA. In the mid-1970s, however, the membership of the LGRB became increasingly dominated by APA critics who were more interested in undermining the agency than influencing it – an ideological shift that paralleled the rise of D'Elia's League for Adirondack Citizens' Rights. In 1976, some of the abolitionist LGRB members objected to certain proposed amendments to the APA Act that the LGRB, led by Purdue, had negotiated. The amendments were designed to alleviate some of the Act's most odious aspects (such as criminal penalties for zoning violations) and, most significant, to make the chairperson of the LGRB a member of the APA board – thus guaranteeing that the agency's leadership had at least one locally chosen voice. The lobbying efforts made by Purdue and other like-minded members of the LGRB were enough to get the amendments through both houses of the legislature. But some abolitionist members of the LGRB lobbied *against* the amendments. These dissidents said the proposed changes were the products of appeasement, and began organizing "speakouts" at which they and others (most of them members of D'Elia's League for Adirondack Citizens' Rights) demonized Governor Hugh Carey and the APA. While Carey signed most of the proposed amendments, he responded to the attacks by vetoing the bill that would have appointed the LGRB chairperson to the APA. Dick Purdue left the LGRB soon after, and was succeeded in the chairmanship by D'Elia. The entire episode, Purdue said later, "was a watershed event, where the abolitionists became the loudest local voice."

Like the environmental coalition that had first established the Adirondack Park, the Adirondack property-rights movement drew diverse political constituencies around a common purpose. In this case, the constituencies were developers and large landowners who joined with citizens who felt a sense of personal outrage at a perceived injustice; the common purpose was the abolition of the APA. The Local Government Review Board was merely the "official" manifestation of this movement, which was carried out on other fronts as well.

Throughout the last half of the 1970s, the abolitionists tried various legal and legislative maneuvers to destroy the APA. Their main argument was that the agency's regulations were stringent enough to constitute a "taking" – that is, to violate the Fifth Amendment of the U.S. Constitution, which states that ". . . nor shall private property be taken for public use, without just compensation." The abolitionists also argued that the APA was unconstitutional because it treated residents within the Blue Line differently from those outside it. Though the legal merit of these arguments proved dubious, they were apt expressions of a mounting fury.

The legislative attempts to kill the APA failed, even though the region's state senator, Ron Stafford, was a powerful member of the Republican-controlled New York Senate. Stafford, who had first been elected to the Senate in 1967, represented both the locals enraged by the red-taping of the little guy and those development, landowner, and timber interests that resented the agency's rigid restrictions. While Stafford did not call outright for dissolution of the APA, he reiterated abolitionist grievances that the agency would not listen to locals and that its restrictions caused the region's economic problems. (In reality, the Adirondacks' cycles of high unemployment and other economic troubles during the 1970s – and later during the early 1990s – paralleled those experienced by most other rural areas of the Northeast, including less regulated areas in New Hampshire and Maine. But the coincidence of bad economic times with the emergence of the APA provided critics the opportunity to blame economic woes on the agency.) As the years passed, Stafford became masterful at simultaneously leading and following anti-APA sentiment. He would foment local anger by criticizing the agency in the press one day, then shrug his shoulders the next as he explained that he'd *like* to work with the APA, but couldn't do so when the locals were so clearly against it. This argument effectively handed the political reins to the Local Government Review Board, which was consistently uncooperative.

Ron Stafford eventually became the deputy senate majority leader and the chairman of the senate finance committee. These positions enabled him to squash almost any legislation pertaining to the Adirondacks, and made other senators (and governors as well) solicitous of him. Stafford didn't have enough power to actually overturn the APA law, however, which is why the several legislative attempts to dismantle the APA fell short. Lawsuits seeking to abolish the agency also fizzled, for the courts did not accept the property-rights argument that the APA's regulations amounted to a "taking" in the constitutional sense.

If these groups did not succeed in abolishing the agency, they did manage to make anti-APA sentiment the most potent political force in the park. Like Senator Stafford, local officials quickly learned that it paid to berate the APA – and that to cater to the agency in any way was to be accused of being a traitor, of "negotiating with the enemy." A few local officials got away with cooperating with the agency by convincing their voters that, like it or not, the APA had to be dealt with. But in most cases the surest route to office was to hate the APA more than your competition did.

As their most aggressive assaults on the park agency failed, the abolitionists' momentum slowed. By the 1980s, most Adirondackers came to accept, if grudgingly in some cases, that the APA would remain. Many people ignored the whole issue, for in reality the agency's regulations directly affected only a few Adirondackers at any given time.

The abolitionists fell into a standoff with the APA, continuing to argue that the agency should be done away with, but lacking the resources and openings to make fresh attacks. Many local government officials simply stonewalled, refusing to cooperate with the agency or to create APA-approved local zoning plans, even though creating such plans would have given them some of the local permit-review control they sought. Many blamed their towns' problems on the APA whenever possible. Meanwhile, the property-rights activists, entrenched in their hatred of the agency, would mount occasional skirmishes: a rally here, a letter-writing campaign there. But for the most part the conflict fell into a stalemate – a cold war that soured chances for progress, but which, like the calm over a Kansas missile field, could be mistaken by the unwary for peace.

Then in the mid- and late 1980s came the Northeast real estate boom. After many lean years, Adirondack contractors were building record numbers of vacation homes and condos. The park agency reviewed three times as many subdivision applications in 1989 as in 1984 – yet estimated that it was reviewing only about half the subdivisions being built. (Most developments in hamlets did not require review, and some elsewhere were being built in defiance of permit requirements.) There was unprecedented building activity not only in large subdivisions, but in the many small projects – those of one, two, or a half-dozen homes – that posed the threat of "a thousand cuts." Many projects large and small were concentrated along shorelines, which had been left weakly protected by the compromises made in the original APA legislation. And as elsewhere in the Northern Forest, speculators were buying big tracts of land and dividing them up into "wilderness lots" of the allowed minimum size (in this case, 42.8 acres), fragmenting the landscape. All of these trends worked their way through the weak spots in the APA's regulations.

As more subdivisions and houses rose among the trees, New York environmentalists began worrying that this surge in development would change the Adirondacks forever. Then the 1988 Diamond sale, which put 96,000 acres of Adirondack forestland in the hands of a developer from Atlanta, Georgia, sharpened concern to a fine edge. This was when Congress and the governors of the Northern Forest region authorized the four-state Northern Forest Lands Study (NFLS), the predecessor to the Northern Forest Lands Council (NFLC). Yet the NFLS, like the NFLC after it, had no regulatory powers; the APA did. New York environmental groups took the obvious course: They began pressing Governor Mario Cuomo to strengthen the APA's regulatory powers to stall development, and quickly.

Cuomo did as Rockefeller had: He appointed a blue-ribbon panel, the Commission on the Adirondacks in the Twenty-First Century, generally known as the Cuomo Commission. In so doing, he unwittingly heated the cold war into a full-fledged confrontation. The two sides, environmentalists versus property-rights activists and some local officials, would feud so bitterly that they would not only negate the Cuomo Commission's efforts, but would taint the subsequent work of the Northern Forest Lands Council as well, producing the anger and fear so evident in Carol La Grasse's tirade at the NFLC four years later.

Opinions differ on what went wrong with the Cuomo Commission, but as with the Rockefeller Commission, it seems easy in retrospect to spot the primary mistake: In choosing the

Commission, Cuomo and his staff failed to include meaningful local participation. By repeating this historic error, they set the cause of Adirondack protection back at least five years, if not much longer.

The failure to include locals on the Commission began right at the top, as Cuomo appointed Peter Berle, the president of the National Audubon Society, as chairman. Berle's distinguished list of accomplishments and positions may have made him an obvious choice for environmentalists, but to Adirondackers resentful of APA authority, the same record made his name poison. As a state assemblyman from Manhattan in 1971, Berle had led the legislative effort to pass the original APA act. During the 1970s he served as director of the New York State Department of Environmental Conservation (DEC), which helps manage state lands in the park and has review authority on some projects; after the APA, the DEC is the most hated bureaucracy in the Adirondacks. And Berle's position at the National Audubon Society didn't stir much confidence in people who felt environmental groups already had too much say in their lives. His Manhattan address didn't help either. Thus Berle's appointment, while pleasing advocates for stronger land protection, suggested to many local Adirondackers that critical decisions about their communities would once again be made by outsiders. This fear was reinforced by the make-up of the rest of the Cuomo Commission, which was dominated by nonresidents and environmentalists. Cuomo did put a few local residents on the Commission, but they were drawn largely from the Adirondack Council, a local conservation group that was despised by hard-line property-rights advocates nearly as much as the APA itself. The Commission's research staff had similar backgrounds and was headed by George Davis, who had been a staff ecologist for the original Rockefeller Commission, had served on the APA board in the mid-1970s, and had since become a leader of the Wilderness Society.

These appointments, made in January 1989, drew only a quiet reaction at first, just as the establishment of the Rockefeller Commission had twenty-one years earlier. Even those such as Senator Stafford who held themselves as champions of local representation failed to object strongly. But there were ominous grumblings. The *Adirondack Daily Enterprise,* the region's only daily, which had shown a clear anti-APA sentiment since the mid-1970s, offered what was perhaps the most cogent protest:

> The Governor's charge to the [Commission] makes clear that he considers the private title to private property in the Adirondacks to be substantially different than the same title on a similar piece of property in . . . Long Island or New York City . . . The staff will be the same old folks from the environmental lobbies and the state departments already managing the public and semi-public land in the Adirondacks (semi-public land is what used to be considered private land) . . . The leader of the group, from Manhattan, will try to reassure the full-time residents of the Adirondacks that he (and the commission) will listen to their concerns and will be 'sensitive' to them. In reality the group will be as sensitive as was Gov. Cuomo when he decided to load up the commission with people from outside the Adirondacks and to shun any person elected by people from the Adirondacks.
>
> There will probably be 'public hearings' and forums. What lies ahead is the fraud of public participation. Cuomo has demonstrated in his choices that any legitimate voice of the residents here will not be included.

After this initial outcry, things were fairly quiet until September 1989. Then the commission released its interim report, which stated that there was a clear development crisis in the park that needed comprehensive solutions. The anti-APA constituency became aroused as a broadening of the agency's powers began to seem imminent. As the commission worked in increasing secrecy on its final report during the following winter and spring, several vocal opponents emerged.

287

The most notable antagonist was Don Gerdts, a former Long Island advertising executive who had moved to the Adirondacks in the mid-1980s and bought a thousand acres, planning to develop it. Gerdts had developed around 10 percent of this land by the time the interim report came out; the rest he'd decided to sell because the real estate market had cooled. A master of the sound bite and the incendiary phrase, Gerdts directed a steady, well-aimed stream of bile at the Cuomo Commission. This commission, he said, was yet another bunch of elite, effete enviros from downstate who didn't care a whit about real Adirondackers and in fact wanted to get *rid* of them – run them out of the park – so the greenies could enjoy the place in peace. Gerdts resurrected all the American Revolution slogans about taxation without representation, colonization, and serfdom. He railed against "King Cuomo."

None of this was new rhetoric. But Gerdts delivered it with the fervor of the newly converted, the craft of the advertising executive he had been, and the indignation of the frustrated developer he was. He founded the "Citizens Group of the Adirondacks" and held rallies at which he inspired audiences to near-violent frenzies. At one rally, held two months before the release of the final proposals, he inflamed the crowd by predicting that the Cuomo Commission would call for a moratorium on building, tougher zoning requirements, and new public lands – thus evoking the ultimate bogie, the specter of purchase by eminent domain. Gerdts did more than anyone to polarize the situation, and for that he may be roundly blamed. But the response he received confirmed that the bitter opposition that had threatened the APA's existence in the 1970s had not disappeared, as some may have thought, but had simply lain dormant, waiting to be aroused.

Indeed, the backlash revivified the coalition that had tried to abolish the APA in the 1970s, an angry alliance of landowners, developers, citizens groups, local government officials, and Senator Ron Stafford. The citizens' groups conducted a vigorous letter-writing campaign and held repeated rallies, including a Memorial

Day "slowdown" caravan on Interstate 87 (the main path from New York City to the Adirondacks) in which hundreds of cars, most of them with signs berating the APA, drove ten miles per hour below the speed limit, backing up traffic for miles. Local officials, arguing that the Cuomo Commission was so unrepresentative of local concerns as to be illegitimate, formed the Association of Adirondack Towns and Villages (AATV) and claimed that the AATV, not the APA, should make land-use decisions in the region. The AATV began echoing the citizens' groups charges that any town official who talked with the commission, the APA, or the governor was negotiating with the enemy. And Senator Stafford, though he had made only token public objections to the Cuomo Commission when it was formed, now questioned the commission's legitimacy. Stafford further reinforced the gridlock by saying he wouldn't back any proposals or representative bodies that weren't supported by the locals – by which he essentially meant the Local Government Review Board and the AATV. In the heat of all this, the LGRB voted not to meet with the Cuomo Commission as it shaped its final report.

Some local newspapers, led by the *Adirondack Daily Enterprise,* also accused the Cuomo Commission of ignoring local wishes – an accusation encouraged by the clumsy way the commission had raised some of the issues in its interim report, and by its increasing secrecy during the drafting of final recommendations in the winter and spring of 1990. The *Daily Enterprise's* editorial writers, for instance, found fertile material in the interim report's objections to rock concerts and wrestling matches staged in the Lake Placid Olympic Village. "Well, excuse us for living," intoned the paper. "Perhaps everyone who calls the Park home, as opposed to a vacation resort, should clean their cages so as not to offend visitors to New York State's six-million-acre petting zoo."

Into this atmosphere the Cuomo Commission released its final report in May 1990. Here was the fan; the feces hit it; and the resulting spray buried any hope that the report's 245 recommendations would be considered calmly. The spray also stained the Northern Forest Lands Study, which had been released the previous month, as New York property-rights activists aroused similar groups in other Northern Forest states with dire warnings about "greenlining" and "land grabs."

Most of the Cuomo Commission's recommendations were concerned with one or more of five principal goals: restructuring rural-use and backcountry zoning to prevent further forest fragmentation; controlling shoreline development; boosting local economies through the creation of a community development bank and through grants to towns for sewage and landfill improvements; expanding both the Blue Line and (by 600,000 acres) the "forever wild" forest preserve; and establishing a more powerful "Adirondack Park Administration" with broader regulatory powers. Both the land purchases and the economic aid would be funded through an Environmental Bond Act already proposed separately. To protect "regional character," the recommendations contained a few proposals concerning aesthetics. Some of these were sensible enough, such as those providing guidelines for signage, but others were hare-brained and insulting, such as one requiring all roofs to be of certain specified colors. The same day the commission released the report, it sent to the legislature an omnibus bill containing almost all of its recommendations – a move that critics said showed both the commission's arrogance and its eagerness to bypass local people.

Not surprisingly, public attention focused on the proposals that best illustrated the outsider-versus-local conflict that had been established over the previous twenty years and intensified by the fallout over the commission's membership and interim report. Don Gerdts, local government officials, representatives of timber and development interests, and Senator Stafford all fixated on proposals that seemed to confirm fears they themselves had helped elicit. In particular, the commission's critics objected to proposals calling for a one-year moratorium on all development, the extension of the forest preserve and Blue Line boundaries, and the expansion of the Adirondack Park Agency's size and authority. Local and regional papers reported these criticisms widely, and many editorial writers echoed them. The outcry redoubled when one of the few local members of the commission, Robert Flack – a former town official, former APA commissioner, and a director of the paper company Finch-Pruyn, one of the Adirondack's largest landowners – released a "minority" report raising many of the same complaints. Underlying all these objections, and present in virtually every news story and press release about local reaction to the report, was the lament that the proposals were cooked up by outsiders behind closed doors, with no meaningful input from locals.

The backlash grew broader. The month the report was released, a Crown Point couple named Dale and Jeris French founded the Adirondack Solidarity Alliance, which would become an even more effective APA antagonist than Gerdts's group. Dale, a former nuclear power plant engineer, was a builder of log homes; Jeris ran a roadside gift shop on the eastern edge of the park. The gift shop soon became Solidarity's office. Solidarity worked closely with Gerdts's Citizens Group of the Adirondacks for a few months, but by the middle of 1990 split off as the Frenches grew leery of Gerdts's increasing shrillness. Less openly provocative than Gerdts, Solidarity claimed to be a relative moderate in the debate, and the Frenches' public rhetoric was less bombastic than Gerdts's. Yet the Frenches were clearly angry, and like Gerdts they replayed the old slogans about taxation without representation and colonialism, and muttered regretfully about being unable to control the desperate souls being driven toward violence.

The references to violence were taken more and more seriously. Rumors spread of armed men meeting in the woods. At one point there *were* armed men in the woods, when a group of rifle-wielding citizens ended up in a tense stand-off with state officers attempting to close a road that had recently been declared part of a wilderness area; the officers backed down. At a later property-rights demonstration on the same road, a town supervisor assaulted one of several Earth First! activists who had come to protest the protest. At a rally for Senator Stafford on the statehouse steps in Albany, a Solidarity member punched John Sheehan, a spokesman for the Adirondack Council, in the face; Sheehan punched the man back. More serious, an anonymous group calling itself the "Adirondack Liberators" wrote letters to editors threatening to torch the Adirondacks' wild forests and the homes of environmentalists. APA staff members received threatening calls at their homes. The APA and Adirondack Council offices were repeatedly vandalized and harassed – sprayed with paint, liquid cow manure, rocks, and epithets. Someone burned down an APA board member's barn.

Governor Cuomo saw that he needed to quickly quell this backlash if he was to pass any of the commission's proposals. Within weeks of the report's release he had eliminated the building moratorium, distanced himself from the omnibus bill and its most controversial measures, and begun meeting with Adirondack town officials. Many of these local leaders spurned him. But a few stepped up to try to create some meaningful negotiations between APA critics and park advocates.

Most prominent among the conciliators was Dick Purdue, a small-scale builder-developer and the Indian Lake town supervisor (a post equivalent to mayor) since 1980. It had been Purdue who, as head of the Local Government Review Board in 1976, had been driven out by the abolitionist faction within the LGRB. A graduate of Princeton and Columbia, Purdue had experience as both a foreign diplomat and as mayor of Ossining, New York, before moving to Indian Lake in 1969 and establishing a small home-building business with his wife. Purdue was a highly independent thinker who defied the stereotype of the Adirondack local official. It wasn't that he was uncritical of the APA. Indeed, he had long criticized the agency for excluding locals, pressed it to be more reasonable in its permit reviews, and excoriated it in the press whenever it made one of its insensitive blunders. But unlike the abolitionists, Purdue believed that the APA's problem was not that it existed, but that it lacked meaningful local representation. He stressed the need for local participation not as an inflammatory complaint, but as a goal to be achieved. He considered the "don't negotiate with the enemy" stand to be an artless dodge that made progress impossible. As a result, he sometimes criticized other local officials or Senator Stafford for failing to influence the APA when they had a chance.

In the late 1980s, Purdue became active in the Intercounty Legislative Committee (ILC), a group of about forty local officials that met monthly during the state legislative session to discuss Adirondack concerns with state legislators. In late 1989, when the Cuomo Commission released its controversial interim report and was rebuffed in its subsequent overtures to the Local Government Review Board, Purdue was among those on the ILC who formed a "Planning Commission" to consult with the Cuomo Commission as it drafted its final recommendations.

In its first few months, the ILC didn't make much progress in its effort to work with the Cuomo Commission, for this was the period during which the commission, stung by the backlash from its September 1989 interim report, worked in increasing isolation. When the final report was released in May of 1990 and Cuomo was subsequently shunned by the Local Government Review Board, he turned instead to Purdue's ILC Planning Commission to try to establish some dialogue.

In the explosive atmosphere surrounding the report's release, points of accord proved elusive. Initially, the ILC Planning Commission wanted to talk only about membership (they were insisting that the governor agree to pick five APA board members from a list supplied by the Intercounty Legislative Committee), and the governor's staff wanted to talk about everything else.

"We started off fencing, trying to make points," says Purdue. "But as we got to know each other, we dropped our unmovable agendas and started to talk about other things in a more concrete and objective way. We talked for over two years, and over that time we moved steadily away from posturing and started to talk rationally about things."

Purdue was further encouraged when the Planning Commission and Cuomo staffers were joined by representatives from some of the area's environmental organizations. This diverse working group seemed to present an historic opportunity to reconcile the APA's mandate to protect the park's environment with residents' desires to have a real hand in the regulations that affected them.

"We found a lot of common ground," says Purdue. "We came to significant, working agreement about a number of things. Land-use control. Water quality. Prospective overdevelopment of shore-front. The loss of open space and the possible decline of forest production. We also shared the state's more reasonable concerns about roadsides and scenic vistas. We got to the point where we were talking about ways to deal with those problems that would be acceptable to the people of the Adirondacks. And we talked in more concrete terms every time we met. We were really getting into the nitty gritty." The working group eventually agreed to create a Citizens Advisory Council, chosen jointly by county government and the governor and installed by law, which would have considerable influence over the APA. In short, they seemed to be close to resolving some of the most recalcitrant sticking points between the APA and its critics.

It was at this point, says Purdue, in late 1991, that a combination of power politics and the old polarization between environmentalists and abolitionists destroyed all the work that the ILC Planning Commission, Cuomo's staff, and the other conciliators had done. For reasons that aren't clear, Cuomo grew impatient. Instead of giving his staff, the ILC Planning Commission, and the participating environmental groups more time to produce consensus legislation, the governor offered a bill of his own in the fall – even though the bill wouldn't be considered until spring.

Cuomo's legislation was far less ambitious than the original Twenty-First Century Commission bill, avoiding the earlier bill's worst blunders and most onerous restrictions, and including some important economic measures that would help towns. But the Cuomo bill had several provisions Adirondackers found objectionable – only two APA board positions for locally nominated members; shorefront zoning measures that usurped some of the towns' existing permit review authority; and other provisions, such as one requiring a two-thirds majority of the APA board to make any changes to the zoning map, that seemed calculated to minimize local influence. Perhaps most significant, Cuomo's bill did not reflect the understandings reached by the ILC Planning Commission and the Governor's staff.

The effect of the Cuomo bill was to return the question of how to protect the Adirondacks to the same polarized political process that had long proved unfruitful. The environmentalists started lobbying for the bill, the abolitionists against it; Stafford again held the keys to the gate. The Planning Commission was left out in the cold.

It soon became apparent that Cuomo saw the bill as merely an opening gambit. "The governor's staff told us in very strong terms that the governor was willing to back away from the bill," says Purdue. "He *wanted* to back away from it." Purdue believes that Cuomo, confident that he knew from the Planning Commission what Adirondackers would accept, had deliberately "overwritten" the bill – that is, that Cuomo drafted the bill with language to please environmentalists while intending to offer Stafford enough concessions that the final bill would reflect those agreements which the governor's staff had worked out with the Planning Commission.

If this was Cuomo's intention, he miscalculated, for Stafford showed no interest in negotiating. Instead, buttressed by his Republican Senate majority, he killed the bill. In the meantime,

Cuomo's end run around the ILC Planning Commission left that body with little credibility with either Stafford (who had previously had little choice but to defer to their "local wishes") or with their broader constituency.

Purdue resented this bitterly. "In withholding his concessions – concessions the Planning Commission had won – to save them as bargaining chips with Senator Stafford," he says, "the governor showed tremendous disrespect for real Adirondackers. We negotiated long and hard to get these things, then they were taken off our plate and fed to Senator Stafford. But he wouldn't eat. There were no concrete negotiations of any sort. The governor was left stuck with his unpassable program bill, and the Planning Commission could not even claim the victories for the many understandings we had worked out. It just cut the legs right off us. It completely destroyed our credibility.

"I can't overcome that kind of crap. I believe in what the agency is trying to do. I want to preserve this environment. But for God's sake, you *have* to treat Adirondackers like they are real people. You can't drop them out at the last minute as if you're playing games with them."

In the following months, the once-constructive relationship between the ILC Planning Commission and the state turned sour. This was due partly to the ILC's resentment about Cuomo's bill, and partly to a change in the group's membership. Two key members retired, and two others were run out because they were discovered to have "ties" to the environmentalist Adirondack Council. (One member's wife was a "dues-paying" member of the Council; the other had a daughter who worked there part-time.) All four were replaced by more reactionary representatives. And Purdue became the center of another controversy when he publicly chastised Stafford for not negotiating on Cuomo's bill. Some in the ILC tried to oust Purdue as chair of the Planning Commission. The bylaws gave only the ILC's Planning Commission that authority, however, and the Planning Commission voted to retain him.

Nonetheless, the Planning Commission's effectiveness was shot, and the larger atmosphere became more antagonistic than ever. Cuomo's legislation failed; the Planning Commission bickered, stopped meeting, and ultimately disbanded; all returned to their trenches.

"And this is going to hurt," says Purdue. "Not just the environment, but Adirondackers. Of course, water quality, roadsides, scenic vistas, and other resources of the park are going to suffer. And if Adirondackers don't find a way to cooperate, we also miss out on things the state can help us with, like getting another Adirondack community college. Getting a custom-tailored economic development program. A school-aid formula that reflects our particular problems in education. You can go on and on with the list of things we could have gotten if we had come into the real world and stopped sucking our thumbs in the corner. We're losing a lot by not growing up and entering the political process."

The great, painful irony of the whole situation, says Purdue, is that most local Adirondackers don't find these issues impossibly divisive. "That's why I can keep getting elected – I won again in early 1994 – even though I'm completely at odds with the policies of just about every other Adirondack official. My voters judge me by local matters, and to the extent they follow me on environmental matters, they figure I'm trying to do the right thing. When presented with environmental issues in a fair way, people favor environmental protection. Adirondackers don't have an ax to grind against the environment. It's ridiculous to think they should! But this layer at the top, this scum at the top of the pond, has managed to convey that message. You get below that, the average Adirondacker is not that far from me. Yet he never gets the chance, because he never gets the issues presented to him in a way that gives him a free choice."

And for this, Purdue puts equal blame on local leaders, the state, and on environmentalists who fail to work with locals.

"If the state and the environmentalists were willing to give residents more credence, you might actually get somewhere. It's not enough just to sit down at the table with them; you have to treat them as equals. That's the main problem. And that can be cured at any time."

20

Slippery Ground

In the cramped office of the Northern Forest Lands Council, executive director Charles Levesque sits in front of a cluttered desk replacing the cover on a personal computer. He fits the gray plastic casing into place, leans onto one hip, and digs into the pocket of his trousers for a half dozen tiny screws and a small screwdriver. "This should only take a minute," he says. He guides the first screw through a hole in the cover and tightens it with four quick turns.

The renovated farmhouse that contains the Council's office is part of a long, low building called the Conservation Center, high on a wooded bluff above the Merrimack River on the outskirts of Concord, New Hampshire. Outside, the trees are bare; it is late February. Inside, Levesque and his two-person staff work in three small rooms that give an impression of chaos barely controlled. Every inch of space has been put to use. Sheets of easel paper bright with magic-markered lists are tacked on the yellowed plaster walls, rolled up in the corners, splayed across a desk. Rough pine shelves on metal brackets sag under rows of three-ring binders and spiral-bound reports. Maps mounted on sheets of foam core lean against the walls. A photocopier blocks the bricked-up fireplace; the mantel, like every horizontal surface, is stacked with boxes of letterhead and piles of newsletters. On a bulletin board near the door are pinned seventeen blue and white plastic name tags, one for each member of the four-state Council. Without looking up from his task, Levesque explains that he collects the tags after each Council meeting. He wants to be sure they will be available for the next.

299

As the Council's executive director, Levesque is part troop master, part administrator, part referee. Some days he is also part computer repairman. The Council's long-awaited draft recommendations are due back from the printer in a week – on March 2, 1994 – just twenty-four hours before news conferences in the four states will release the document to the public. Levesque knows that the staff will be fielding and responding to hundreds of comments and questions in the months ahead, and this computer, which is run by administrative assistant Mary Beth Hybsch, needed a new disk drive to handle the load. So he put one in.

Levesque, who is thirty-six, first came to the Conservation Center in 1983 to work for the New Hampshire Timberland Owners Association, one of several organizations housed here. Before that he had been a forester for Brown Company in Berlin, for the U.S. Forest Service in Montana, and for the city of Manchester, New Hampshire, his hometown. As director of the Timberland Owners Association in the mid-1980s, Levesque lobbied hard for the forest industry in the legislature and the press. In 1987 he left the group (but not the building) to work for the Trust for New Hampshire Lands, a nonprofit group formed to administer a $47 million public land conservation program. There he advocated for public land acquisition as energetically as he previously had for private forest management.

Levesque was at the Trust in 1988 when 970,000 acres of former Diamond International land were sold in the Northern Forest, and he was involved in the state's purchase of 39,500 of those acres in northern New Hampshire. After a study by the U.S. Forest Service and the four-state Governors Task Force on Northern Forest Lands concluded that the Diamond land sale indicated disturbing trends, Congress and the governors of Maine, New Hampshire, Vermont, and New York established the Northern Forest Lands Council to continue the research and develop a conservation plan. Established and funded through the 1990 U.S. Farm Bill, the Council was originally made up of twelve governor-appointed members – landowner, environmental, and state agency representatives from each of the four affected states – and one official from the U.S. Forest Service. Later, four members representing local interests were added. Each state's forestry department assigned at least one staff coordinator to the project. Levesque fit in well as chief of staff to this diverse coalition. His experience in timber and land conservation satisfied the industry and environmental delegates, and his administrative and lobbying skills appealed to the state agency heads.

Nothing, however, had prepared Levesque for the high visibility and political complexity of the debate the Council was supposed to mediate. By the time the group held its first meetings in early 1991, the uncertain future of the 26-million-acre Northern Forest had become the Northeast's most confounding land conservation issue. Environmentalists were pushing for tough measures to protect wildlife and scenery, the forest industry was trying to squelch talk of regionally enforced logging regulations and federal land acquisition, residents of the affected areas were increasingly nervous about distant entities discussing decisions that affected their lives, and indignant property-rights groups were swelling out of the Adirondacks.

In attempting to shape these conflicting agendas and fears into a workable conservation plan, the Council was determined not to repeat the mistake made by Mario Cuomo's Commission on the Adirondacks in the twenty-first Century, the mistake of alienating the region's residents. Levesque, resource specialist Esther Cowles, and the state coordinators worked hard to involve as many people as possible in the Council's deliberations. Virtually all meetings were open and held at convenient places and times. Anyone who asked was appointed to one of the Council's working groups or Citizen Advisory Committees, which included individuals from across the political and socio-economic spectrum. Thousands of people attended conferences and "public input sessions" held throughout the Council's three-year life. The state coordinators compiled a natural resource inventory and a high-tech digitized map of the region, and the Council's subcommittees commissioned eighteen separate studies on such topics as biological diversity, forest-based economic development, and tax policy. Every Council document was available for public review, and the staff dutifully collated all public comments and distributed them widely. The mailing list for major documents and for the group's quarterly newsletter eventually grew to over thirteen thousand names.

After two years, in September 1993, the Council summarized its research and listed more than two hundred possible recommendations in a document called "Findings and Options." This report was a trial balloon – a squadron of trial balloons, actually – to see which ideas would float and which would drop. It elicited more than a thousand pages of public comment. When the Council met in a two-day closed-door session that fall, the members, sometimes arguing fiercely, narrowed the proposals down to thirty-three that had broad public support. It took the staff and the state coordinators another three months to draft language with which all the Council members could abide. During that time Charlie Levesque and Esther Cowles, along with principal writer Charles Johnson of Vermont, had been working sixty-hour weeks.

Levesque tightens the last screw on the computer cover. He plugs the power cord into an extension outlet on the desk, places the monitor back on top of the machine, and flips the "on" switch. After a moment, a reassuring series of gray characters flash on the screen. He nods, satisfied. He steps out from behind the desk, leads the way into his private office, and settles into a creaky vinyl chair. The wall behind his chair is covered with a map of New England and New York.

"We're frazzled," Levesque says, running a hand through his thinning dark hair. In the eight days between now and the public release of the report, he has to organize a road trip to Washington to present the plan to the states' Congressional delegations, choreograph the news conferences, oversee the mailing of the report, and explain to a constant stream of callers why they can't get early copies. He smiles thinly and raises his hands to mimic an archer aiming and releasing an arrow.

"The property-rights folks are chomping at the bit to let fly at this thing," he says. "They're networked all across the region. We've been trying to do one-on-one meetings with them and it's just not working. No matter what we say, they say 'Yes, we know all that, but what do you really want? What's your hidden agenda?'"

The Council has been confronted by these questions from the beginning. After tense encounters with right-wing property activists in New York and Maine in 1991, Levesque had identified the angriest and most vocal individuals, including several veterans of the conflicts in the Adirondacks, and brought Council members to their homes for private talks. He showed them the Council's carefully worded mission statement and its "findings," and gave them biographical sketches of the members. These visits soothed the fears of some critics, but not all. In Vermont, for example, Levesque and Council members Conrad Motyka and Richard Carbonetti, both of whom lived in Vermont and were quite moderate on land-use issues, visited Tom Morse of the Vermont Property Rights Center. Morse (who had once hung Vermont's moderate Democratic governor Howard Dean in effigy outside the property rights group's headquarters) had been saying that the Council was a preservationist group advocating a twenty-six million-acre national park in the Northern Forest – one of the standard property-rights accusations against the Council. Levesque and the two Council members spent two hours explaining that the Council was purely advisory, that the group had rejected the idea of recommending large land purchases, that some of its members were quite conservative and defended property rights vehemently, and that the options under consideration would benefit landowners in several ways. But all Morse could see was the Council's endorsement of modest public acquisitions.

"At the end he said, 'You don't understand,'" Levesque recalls. "'Up here this Council is out of control.' Carbonetti says, 'No. We are the Council up here.' But Morse just didn't get it.

"Another guy said his bottom line was simple. If the Council didn't recommend cutting all government by 75 percent he wouldn't be satisfied. End of discussion." Levesque pulls a letter from a loose pile of papers on his desk. "Take a look at this."

The letter, addressed to the Northern Forest Lands Council, is from Stephen Stange of Lincoln, Maine. "Re: Criminal behavior," Stange's letter begins. "I become sick to my stomach every time I read one of your publications. It seems you are implying that anyone who does not share your perverted, socialistic, land rights grabbing agenda does not 'care.' What dribble. Perhaps you feel that I am not serious about wanting to see you and your criminal buddies in prison, but I assure you I am serious. Get the hell out of Maine."

Levesque sighs, and replaces the letter in the pile. Ideologues like Stange and Morse represent only a tiny minority, but their accusations of a green conspiracy are part of a crossfire that at times has made Levesque's job nearly impossible. For while the far right has accused the Council of doing the environmentalists' bidding, many environmentalists have made the opposite argument: that the Council was caving in to timber companies and landowners.

That accusation dates back to 1989, when the Governors Task Force was struggling to define the role of the yet-to-be-created Northern Forest Lands Council. Ted Johnston, president of the Maine Forest Products Council, had threatened to withdraw the support of Maine's forest product industry if the new Council were given authority to pursue regional land-use controls or large-scale public land acquisition. It was a potent threat. Maine's big forest products companies owned 10.8 million of the 26 million acres in question, and the industry held such influence in Maine politics that if the companies withdrew their support, the Council would probably lose the backing of the state's political leaders, including U.S. Senate Majority Leader George Mitchell. In short, the industry held sway over almost half the Northern Forest's land and some of its most powerful politicians.

To a large extent, the ultimatum worked. "Recognizing the necessity of keeping the four-state coalition alive in order to gain federal support, and the simple fact that most of the land area of the Northern Forest was in Maine, the Governors Task Force agreed to a proposal from the Maine delegation for a scaled-down version of the Council," Carl Reidel, a professor at the University of Vermont and a member of the task force, later wrote. "The conservation groups and moderate state officials were no match for the forest industry's political infighting skills." Nor were the moderates prepared in 1991 when property-rights advocates protested early drafts of the Northern Forest Lands Act, a Congressional bill that would have written the Council's duties into law. At raucous public hearings in Maine and Vermont that were chaired by Senator Mitchell and U.S. Senator Patrick Leahy of Vermont, property-rights activists bearing signs such as "NORTHERN FOREST LANDS COUNCIL PUKE" demanded not only that the bill be withdrawn, but that lawmakers abolish the Council altogether. The discouraged lawmakers allowed the legislation to die before it was ever introduced in Congress. From then on, the Council was a private, not-for-profit corporation working with funds granted through the U.S. Forest Service.

In the aftermath of these challenges to its authority, the Council adopted as its mission statement the innocuous language of a 1988 letter in which Leahy and U.S. Senator Warren Rudman of New Hampshire had launched the original Northern Forest Lands Study: "The current land ownership and management patterns have served the people and forests of the region well," the Senators had written. "We are seeking reinforcement rather than replacement of the patterns of ownership and use that have characterized these lands." Many environmentalists complained that the statement was illogical. The current "patterns" of ownership, they pointed out, were precisely what had led to the problematic "uses"

– over-cutting, land conversion, and loss of public access – that the Council was supposed to address. The demise of the Northern Forest Lands Act and the Council's vague mission statement seemed to preclude two measures considered by many environmentalists to be the best ways to solve this dilemma: significant acquisition of public land, especially for ecological reserves; and regulation of intensive timber harvesting practices such as clearcutting.

Ever since then, many environmentalists had chastised the Council for not being bold enough. Few of these environmentalists, however, had pushed their case with the vehemence of the property-rights advocates. This was partially because the statewide conservation groups – the independent Audubon Societies of Maine and New Hampshire, the Society for the Protection of New Hampshire Forests, the Adirondack Council in New York, and the Natural Resource Councils of Maine and Vermont – have long preferred consensus over confrontation. In the past, these groups may have had sharp disagreements with landowners and industry (and in the Adirondacks, with local officials), but they generally had worked through the conflicts without extended legal battles or lasting enmity. Both the Society for the Protection of New Hampshire Forests and the Vermont Natural Resources Council, for example, were sponsors of Tree Farm, a national forestry education and promotion program, and most of the region's groups had timber officials on their boards and foresters on their staffs. Further, once the Northern Forest issue started to attract national attention, environmental groups in the region had formed a loose network called the Northern Forest Alliance, organized under the auspices of the Appalachian Mountain Club, which had helped assimilate the large national organizations – the National Audubon Society, the Wilderness Society, and the Sierra Club – into the debate. And while the environmentalists had avidly promoted conventional protective measures such as public parks and controls on logging, many also recognized the legitimacy of the working forest and had begun to address the need for forest-based economic development. Starting in fall 1993, the Northern Forest Alliance promoted a three-part "vision" of protected lands, carefully managed forests, and healthy local economies.

The environmental community was not a seamless entity, however, and some groups, including Restore: The North Woods, had pushed a more aggressive green agenda than others. Among the most outspoken were contributors to a quarterly newspaper called the *Northern Forest Forum,* published and edited since 1992 by Jamie Sayen of North Stratford, New Hampshire. Sayen and his regular contributors, including author Mitch Lansky of Maine, argued persistently that logging practices posed as big a threat to the Northern Forest as careless development did, and they constantly and unapologetically called for rigid forestry laws and public reserves of up to ten million acres. Jamie Sayen in particular claimed that many aspects of the status quo – especially absentee ownership of land by corporations – were as bad for the region's people as for its environment, and he stressed the need for locally directed economies. Sayen voiced these arguments repeatedly at meetings of the Northern Forest Lands Council and the New Hampshire Citizens Advisory Committee. At the same time, Sayen opened the pages of the *Forum* to anyone who wanted to disagree with him, and while his proposals drew fire from timber advocates, local officials, and even moderate environmentalists, his willingness to participate in the time-consuming and often frustrating debates eventually earned him grudging respect from most quarters.

The lack of a united hard-line environmental faction with a single agenda underscores the vast complexities in the Northern Forest. The conservation groups within the Northern Forest Alliance have markedly different philosophies and constituencies, from Restore: The North Woods on the left to the National Wildlife Federation, which is supported by many hunters and anglers, on the right. The Alliance often took months to finalize even relatively benign public statements, as each organization brought slightly different perspectives and purposes to the debate.

Proponents of private property rights had similar differences, as moderate landowner advocates frequently distanced themselves from the more militant groups. Within the forest products industry, historic tensions and new differences of opinion strained relations among foresters and between loggers, foresters, landowners, and mills. Sharp disagreements also divided many Northern Forest towns, especially paper mill towns where economic and environmental welfare were controlled by multinational paper companies headquartered far outside the region. When faced with broad questions about land use and economic stability, some residents had more faith in these companies, while others trusted state conservation and planning agencies.

The shuffling of these various objectives, fears, and positions created unexpected schisms and partnerships. Jamie Sayen, having won the respect of some progressive timber officials, occasionally persuaded them to raise prickly questions about the industry that Sayen knew would go nowhere if he himself brought them up. At the Council's first meeting in the Adirondacks, in the town of Ray Brook, industry officials Ted Johnston and Bob Stegemann, both staunch defenders of property owner rights, were among the Council members who had to be escorted by state police through a gauntlet of angry landowner advocates. In New Hampshire, the Appalachian Mountain Club, which had operated a century-old system of hiking trails and huts in the White Mountains and generally got along well with area residents, was sharply criticized by local officials for trying to influence the relicensing of James River Corporation's hydropower dams in Berlin. (One outspoken critic of AMC's intervention in the James River dam issue, city councillor Paul Grenier of Berlin, was also an officer of the local union, which had been involved in bitter fights with the mill over layoffs and pay cuts. Around the same time, the city and James River were threatening one another with lawsuits over a property-tax dispute.) In northern Maine, environmental organizer Sandy Neily, working for the Maine Audubon Society, had to condemn a Restore: The North Woods campaign to bring back the timber wolf before her neighbors would let her into their living rooms.

The fact that the conflict had been more a series of skirmishes between murky and shifting alliances than a clearly defined battlefront reflected the political difficulties of addressing many interrelated environmental, social, and economic problems simultaneously. If the groups sometimes seemed unsure about how to address the "Northern Forest issue," it was partly because the existence of the Council – and the commitment of most interests to support its consensus-building process – helped prevent the problems affecting the region from being reduced to one conflict around which a pitched battle could take place. The long, consensus-based Council process forced all who took part to acknowledge complexities of the sort that were lost in, say, the fight over forests in the Pacific Northwest, or in the Adirondacks.

Rather than arguing about whether large clearcuts or intensive lakeside developments caused the greater environmental damage, for example, and then trying to agree on specific regulations to curb these excesses, the Council and its citizen working groups tried to identify the factors that caused landowners to choose onetime profit-taking over long-term stewardship in the first place. Because this approach concentrated on understanding the fundamental economic forces pressing on the Northern Forest, rather than the ways in which specific individuals and companies reacted to these forces, it reduced finger-pointing and eventually produced agreement on what most of the underlying problems really were. This encouraged the diverse Council membership to consider broad-based solutions intended to solve these root problems, rather than controversial policies likely to alleviate only the most visible symptoms.

Charlie Levesque feels that this wide-ranging approach held distinct advantages. Yet he knew that the results of these deliberations would lack the sense of purpose and rhetorical impact that comes from focusing on a single problem. "There's a real lack of trust in these broad-based incentives, because there is no direct cause and effect," Levesque says. "We have this big block of recommendations, but you can't point to them and say with certainty, 'There, that will do it.'"

As Levesque waits for the recommendations to come out, he worries that the report – which the Council has titled "Finding Common Ground" – will be just bold enough to raise the ire of private-property-rights activists, but not bold enough to satisfy environmentalists. In particular he fears that if highly charged arguments develop over specific problems such as forest practices and creation of new public parks, all the time spent trying to identify and address root causes will be wasted, and the fragile common ground will disappear.

Levesque explains that as far apart as the various parties on the Council seemed to be at first, they found themselves closer than they realized once they got to know each other and acknowledged the complexities of the issues. "I was surprised," he says, leaning forward in his chair. "I thought we were here." He holds his hands about two feet apart over his desk. "But we were really here." He closes his hands until they are six inches apart, then slumps back. The question is, he wonders, Will that be close enough?

For six weeks in the spring of 1994, from the Hilton Hotel in Manhattan to a high-school gym not far from Jim Moffatt's home in northern Vermont, the Northern Forest Lands Council held twenty public "listening sessions." During these hearings, more than eight hundred people gave their opinions on the draft recommendations. Another eight hundred submitted written testimony. Some comments ran for dozens of pages; one read simply, "B+."

Levesque's concerns about angry property-rights critics turned out to be largely unfounded. The hearings were generally civil, as virtually all found something with which they could agree and little that was truly offensive. Even the demonstrators who appeared at a few of the sessions carried blandly worded signs such as A GREEN ENVIRONMENT DEPENDS ON A STRONG ECONOMY and FORESTS FOR THE PEOPLE. Most commenters opened or closed their remarks with compliments to the Council members for their open-door policy and their success at reaching agreement.

The review period marked the height of the Council's obsession with public involvement. Every listening session was both tape-recorded and transcribed by a stenographer, and the transcripts were made available on paper or computer disk to anyone who asked. The staff organized all comments, verbal and written, in a specially tailored computer database for easy analysis.

Critics tended to embrace one of two broad philosophies. Those who believed that economic and environmental stability can be achieved by encouraging private landowners and businesses favored the Council's proposed "carrots": property and income-tax reforms; subsidies for good management; assistance in marketing forest products; technical help and education for foresters, loggers, and manufacturers; and relief from inconsistent or unwieldy regulations on truck weights, manufacturing facilities, and workers' compensation. Those who believed that the "stick" of public policy should be wielded more assertively urged the Council to strengthen its proposals on public land acquisition, biological diversity, land-use ordinances, and the regulation of forest practices. Only a few commentators rejected the report out of hand. Some private-property-rights activists complained that the Council had abused its authority, and a handful of environmentalists said the group had abdicated that authority. About 15 percent supported the proposals exactly as written.

The Council met twice during the summer of 1994 to review the comments and craft final recommendations, which were completed and delivered to Congress and the governors that September. The final report, entitled "Finding Common Ground: Conserving the Northern Forest," varied from the earlier version primarily by the inclusion of four new recommendations, more detail on proposals that were included in the draft, a strengthened chapter on forest practices, and the addition of a section suggesting ways that the states could help implement the plan and continue the dialogue begun by the Council.

While written more clearly than many government reports are, the thirty-seven recommendations did not make for lively reading. But even a simplified account of some of the most prominent proposals conveys a sense of both the possibilities and obstacles in addressing the region's problems.

First, the most promising proposals – those that, if acted on, could help solve some of the more pressing problems in the Northern Forest. The Council wisely proposed:

Changing federal and state estate-tax policies so that families like the Moffatts can pass on forestland without forcing their heirs to sell or overcut to pay inheritance taxes.

Making it easier to protect special natural and recreational areas like Lake Umbagog by increasing money for state and federal land acquisition, establishing a clear process for prioritizing target areas, and providing funds and other support for creative land-protection techniques such as conservation easements.

Increasing funding for the federal Forest Legacy program, which purchases conservation easements on working forestland, thereby removing the temptation to develop productive woodlands.

Reducing landowner costs by lowering capital gains taxes on timber and by stabilizing property taxes; these changes could make it easier for more landowners to pursue the kind of environmentally responsible and economically sensible forestry that Baskahegan Company is able to afford.

Encouraging green certification of paper and wood products from sustainably managed forests; this will reward responsible landowners and shift some of the cost of good forestry to consumers.

Reforming workers' compensation rates and rules, to reduce pressures on logging contractors like Treeline of Maine, pressures that compel loggers to cut hard just to break even.

Increasing direct federal incentives to help small landowners manage their forests for wildlife, recreation, and clean water – a cost-effective way to protect these "non-timber" values.

Expanding research into forest health and ecosystem management to help scientists like Carol Foss develop practical solutions for private and public decision-makers.

Strengthening technical education programs for loggers, foresters, and landowners so they better understand how to minimize the adverse impacts of logging on forested ecosystems.

Finally, the Council's hard-won recommendations to preserve biological diversity by promoting better forestry and creating ecological reserves will help promote a more sensible balance between timber production and biotic integrity.

In other areas, the proposals fell short:

The measures to strengthen and diversify the value-added wood-based manufacturing industry with marketing networks and cooperatives are unoriginal and unimaginative, and rely too heavily on "assistance" from already overworked state economic development agencies. Of the several economic development proposals, only the call for Community Development Financial Institutions, which would help finance locally owned forest businesses, holds real promise to make better use of locally grown wood and curtail the export of raw logs and loss of jobs.

Another poorly addressed area is recreation and tourism, a section that sporting camp owner Dale Wheaton called "completely disappointing." Despite the largely untapped economic potential of recreation in the Northern Forest, the Council offered only two ideas: one aimed at reducing legal liability for private landowners who allow recreational access, and one proposing new taxes on recreational equipment. While both are worthy, the Council provided no analysis of the varying economic and environmental impacts of different kinds of recreation. It also failed to offer specific suggestions for encouraging ecotourism or traditional sporting camps, which can provide quality backcountry recreation without increased development.

Finally, the section on land-use planning did not adequately address the fundamental problem that local ordinances often fail because, as happened when Craftsbury considered the problems of growth, lay regulators are often unable to effectively police their neighbors.

The report's omissions generated as much discussion as what it included. Perhaps the most controversial omission was the Council's failure to suggest explicit reforms of logging regulations in the region. The Council chose to leave this issue to the states, even while acknowledging that "Throughout the opportunities for public comment, people have expressed great concern over current management of the forest. The Council recognizes that many private landowners have been excellent stewards of the land, in many cases for generations . . . However, we are concerned that some landowners and woods operators are employing forest practices that compromise the . . . economic and ecological benefits which the region's forests have traditionally provided."

This was hardly the bold indictment of large clearcuts and high-grading that some environmentalists were seeking. Nonetheless, that the recommendations addressed forest practices at all was a major victory, for the industry had long succeeded in keeping the subject off the table. In the Council's final meetings, however, forest practices was, according to Council members, essentially the *only* issue under discussion. This was due to two factors: the strong expression of public concern over abusive harvesting at the Council's public hearings; and an unexpected change in the Maine delegation: In early June, just before the Council met to revise the draft recommendations, Ted Johnston, who had long led the effort to keep forest practices off the Council's agenda, abruptly left his job as head of the Maine Forest Products Council. His seat on the Northern Forest Lands Council

314

was filled by Roger Milliken, president of Baskahegan Company (therefore Chuck Gadzik's boss) and chairman of the board of the Maine Forest Products Council.

"The Maine delegation was much more positive with Roger there," Council member Paul Bofinger of New Hampshire said. "If not for him, we would never have made the progress we did." Brendan Whittaker, the conservation representative from Vermont, agreed. Another insider said that Bob Stegemann of International Paper, who was the industry representative from New York, "really came around" during this period, backing off from his earlier objection to logging regulations once Milliken made some specific proposals.

The Council did not outline specific logging laws for the region. However, it did endorse nine "Principles of Sustainability" by which good forest management should be judged. (These include maintenance of soil productivity, water quality and wetlands, a healthy balance of age classes, quality and quantity of timber, scenery, protection of unique areas, opportunities for traditional recreation, and sufficient habitat for the full range of native flora and fauna.) That industry leaders agreed that these values must be better accommodated in working forests was an unequivocal call for a new set of priorities. To put these principles into practice, the report recommended that states revise laws, educate loggers and foresters, and appoint forestry "roundtables" of diverse interests to develop specific programs.

Perhaps even more significant were two sentences in the final report's introduction: "In our discussion, time and again we faced a fundamental conflict – between market-driven efficiency that encourages maximum consumption of resources with the least amount of effort in the shortest time, and society's responsibility to provide future generations with the same benefits we enjoy today," the report stated. "We believe that until the roots of this conflict are addressed and the economic rules changed so that markets reward long-term sustainability and recognize the worth of well-functioning natural systems, existing market forces will continue to encourage shorter-term exploitation instead of long-term conservation of the Northern Forest."

The Council warned that it could present only "feasible" measures to deal with the symptoms of this most basic flaw; its members were not, after all, in a position to change economic rules and market forces, and they were constricted to what they could reach consensus on. But the brief paragraph declared clearly that the status quo – the "current patterns" of ownership and use – was not always benign. Change was necessary.

As some critics were quick to point out, most of the Northern Forest Lands Council's final recommendations had been floated five years earlier in the 1989 Northern Forest Lands Study. These critics questioned whether it was worth four years and nearly five million dollars of public money to have the Council come up with the same ideas.

While valid on its face, this criticism ignored that the 1989 study was just that: a government study written by Forest Service employees. The Council's recommendations, on the other hand, were endorsed by seventeen people representing sharply conflicting interests, and had been developed through the participation of hundreds of informed, involved citizens. The ideas these people finally approved may have existed five years earlier, but at that point there was no consensus that they would work.

"You can't imagine how hard it was to get even what we got in the area of biological diversity, forest practices, and new land acquisition," said Council member Paul Bofinger, president of the Society for the Protection of New Hampshire Forests and that state's environmental delegate. "If we hadn't pushed as hard as we did, we'd still be arguing about whether any more public land was even necessary, much less whether we should recommend ecological reserves. Whenever you do something that shapes wide social and economic policy, you always have a low common denominator. Well, the Council had half a dozen of them."

Bofinger and other Council members say this without bitterness. They know that their colleagues did their best to reflect their constituents' views and concerns, a task that was particularly challenging for those delegates who represented industry, and even more so for those who spoke for local people. As Charlie Levesque said, "The local representatives were constantly wondering what their neighbors would say about the recommendations, because they had to defend their involvement on the Council every day in the grocery store. So it brought the other members back to Earth pretty quickly. That's okay; that's what they were intended to do."

These constraints notwithstanding, the Council's long deliberations, its emphasis on consensus, and the resulting recognition of the issues' complexities substantially raised the level of discussion. The dialogue moved beyond conventional disputes over environmental protection, property rights, economic growth, and home rule toward a larger examination of how the different forces at work in the Northern Forest – economic, natural, cultural, and political – might be controlled or encouraged to make the entire system healthier and more sustainable.

Partly by default, partly by design, partly by the dogged insistence of a few individuals both within and outside of the Council, the Council had come to look at the landscape as a whole that is greater than the sum of its biological, social, and economic parts – to consider the complex interactions and relationships, including human pressures from afar, that contribute to environmental health or degradation. Thus while taking heat from some environmentalists for not being bold enough, the Council was in fact exercising in politics an ecological concept that has risen to the top of green agendas across the nation: the concept of ecosystem management.

Many people were disappointed that the Council did not recommend continuing a regional body to further develop this concept. And it is unfortunate that the Council could not give this regional dialogue – which proved so productive in generating ideas and consensus, and which virtually everyone agreed was one of the Council's most significant "products" – a framework in which to continue and progress. Some Council members had argued for establishing such a regional entity, but the idea was lost to the intense desire to protect state jurisdictions, unyielding opposition to anything resembling "greenlining," and, ultimately, the exhaustion of the Council members themselves. When the group disbanded as planned in September 1994, it called for ongoing forums within each state and for interstate coordination of ideas and programs, but pointedly refrained from endorsing any regional body.

Charlie Levesque, for one, thinks that's wise. He believes that the Council took the dialogue as far as it possibly could. "We've used this high-profile approach to the max now," Levesque said. "This is the end of the road. Others are going to have to step in and take it from here or not."

"We're all done," Bofinger echoed. "There's just too much baggage, too many commitments that have been made between the players. But by and large the pieces of the puzzle are still there."

21

From the Ground Up

When the Northern Forest debate began in 1988, much of the timberland in the region had become more valuable as speculative real estate than as working forest. This fundamental change in value inspired the sale and fragmentation of the Diamond lands, the rush of speculative investment, the fears of condos covering the backcountry, and the establishment of the Northern Forest Lands Study and the Northern Forest Lands Council.

In the early 1990s, however, even as the Council was grappling with the implications of this change, the trend began to reverse itself. To begin with, the real estate market collapsed from overspeculation, bank failures, and the 1990 recession. Of the more than five million acres of land that had changed hands in the Northern Forest from 1980 to 1991, approximately a quarter million acres had been subdivided for development. Because half the parcelized land was located on the lakes and rivers, where development has its highest impact on wildlife habitat and water quality, this activity disturbed sportsmen and environmentalists. It did not, however, suggest that unbridled development was spreading through the forest. To many of the people who had become concerned about the Northern Forest, this real estate slowdown seemed a well-timed blessing – an opportunity to strengthen land-use regulations and take other steps to discourage land conversion before the market rebounded.

Even as real estate values leveled off, another trend was emerging, the importance of which was not fully recognized until 1993, as the Council was writing its draft recommendations: The price of timber rose sharply and steadily. This increase was partly a reaction to drops in timber supply from other U.S. sources (particularly the Pacific Northwest) and other parts of the globe, including tropical rain forests and Eastern Europe, where years of overcutting had created shortages. The Northern Forest began to take on new prominence as a major international wood source as buyers both domestic and foreign recognized this forest's productivity and resilience, the quality of its hardwoods and softwoods, and its distinction as one of the few places in the world where wood was growing faster than it was being cut. In the early 1990s, prices for virtually all solid wood products soared, and pulp prices rose as well.

Suddenly the world was coming for the Northern Forest. New England sawmills, already paying near-record prices for locally cut wood, found themselves being outbid by exporters by as much as 20 percent. Factory ships loaded with millions of board feet of unprocessed pine and hemlock logs left wharves in New Hampshire and Maine, bound for China and Japan. Bidding wars escalated the stumpage prices for premium white pine, which doubled in some areas between 1988 and 1994. The rate for high-grade hardwood sawlogs, including red oak, rose by 60 percent or more in the same period. Some sawmills started expanding their capacity; at least one began a series of radio advertisements urging landowners to cash in on the high prices.

For landowners and the forest products industry, these rising wood values present a tremendous opportunity for long-term investment – and for quick profits. Charles Niebling, executive director of the New Hampshire Timberland Owners Association, puts it this way: "Either we can cultivate the renewed interest in the wood of the Northern Forest in a way that ensures that the resource and its value is there in perpetuity. Or we squander it to mismanagement or permanent conversion."

Niebling and others believe that this crucial decision – whether to cultivate or squander the Northern Forest – will be made by the early years of the twenty-first century. Some compare the situation to that of New England's commercial fisheries, which showed signs of serious decline for a decade before collapsing from overfishing in the early 1990s. As in the fishing industry, the decision to nurture or exploit the Northern Forest will be shaped not just by broad public policies, but by choices made by thousands of individuals and hundreds of companies. There are strong pressures pushing in both directions.

Henry Swan, president of Wagner Woodlands in Lyme, New Hampshire, is one timber executive who plans to cultivate the Northern Forest for long-term returns. Wagner and its subsidiaries own or manage for others more than half-a-million acres in northern New England and New York, including 238,000 acres that John Hancock Timber Resource Group purchased in 1993. (Hancock Timber is a timberland investing arm of John Hancock Financial Services. Some of the lands Hancock bought were former Diamond holdings that James River Corporation had acquired in 1988.) Swan seeks investors interested in slow, steady capital appreciation over many years. He says that strong markets, along with patient investors, allow him to practice sustainable forest management rather than heavy, cyclical harvests.

Swan believes that many forest managers fail to capture the full economic value of timberland because they harvest wood too soon or use too much wood for low-grade products. He is particularly bullish about the value of fine-grained, light-colored hardwood species such as ash and hard maple (which are especially popular for furniture in western Europe), and anticipates that markets for these woods will only improve as other sources of hardwoods are overcut. "We're not going to find these species in Indonesia," Swan says. "We're not going to find them in South America. We're not going to find them in Russia. But we have them here.

"Timberland will never be a good investment for someone who wants to get 5 percent a year for five years and then get out," Swan says. "We have to find investors who are willing to sacrifice short-term income for a good long-term total return." Swan also has to find land that has not been overcut or subdivided, which he does by flying over the region with his foresters.

Henry Swan is not alone in his belief that only more patient forest management can fully realize the long-term economic potential of the Northern Forest. Along with the Baskahegan Company in Maine, several other commercial landowners who are committing significant acreage to longer rotations include Seven Islands Land Company, which manages over a million acres in Maine for the heirs of nineteenth-century land baron John Pingree; and the Lyme Timber Company of Lyme, New Hampshire, which owns and manages sixty-six thousand acres throughout the Northern Forest region. Many small landowners such as Jim Moffatt are also trying to manage the woods for long-term growth.

As many a recent clearcut attests, however, higher wood values also increase the temptation to squander the forest's potential by overcutting. Vermont's Larry Brown is but one example of a speculator who buys land and cuts it hard to make the most of short-term markets. In late 1994, a land speculator from New York state purchased just under ten thousand acres of forest in Vermont's Mad River Valley, one of the state's most popular scenic areas, for $392 an acre, and immediately sold to a Canadian logging firm the rights to cut every hardwood tree bigger than 12 inches across and every softwood tree wider than 8. In March 1994, a New Hampshire logging contractor and speculator with a reputation for heavy clearcutting bought 1,731 acres in northern New Hampshire for a reported $800 per acre – an unheard-of price for North Country timberland in a depressed real estate market. The only way to repay the mortgage on this property, which was estimated to be over a million dollars, will be to cut every tree; that's exactly what the logger was doing by the fall of 1994. Experts worry that higher wood prices will accelerate such timber liquidation.

Pressures for quick returns also operate in more subtle ways. Take, for example, the demand for pulpwood and chips, which in the Northern Forest are used primarily to make paper, but also to make manufactured wood products such as waferboard, and to supply wood-fired electric plants. Because these industries demand tremendous quantities of trees of almost any size, companies that own both forestland and mills often cut their forests before the trees reach economic or biological maturity. Woodland managers for these firms may recognize that the trees could be more valuable later, but are often compelled to cut them early so they can avoid paying high prices for wood on the open market, and so they can show acceptable quarterly returns. Thus the "vertical integration" of the paper companies – their ownership of both raw material and manufacturing facilities – partly exempts their wood management from the laws of supply and demand, leading the companies to undervalue their woodlands. As one veteran observer notes, "You have to keep the mill going, and there are a lot of shortcuts taken to do it. No honest person in the paper business would say that doesn't happen. If you move forest management decisions from the mill owners to non-industrial landowners who have a longer view, you eliminate one of the variables that leads to liquidation."

"Non-industrial" commercial landowners like Wagner Woodlands and the Baskahegan Company do not have mills to feed, and are therefore not tempted to cut hard just to keep the plants running. Timber officials such as Niebling and Swan believe that having more land owned and managed by such non-industrial companies would help stabilize and strengthen the region's forest-based economy while protecting ecological and recreational values. Many environmentalists agree.

Paper company lands, of course, are not just fiber farms to feed company mills; they are integral parts of the forest products web, selling on the open market everything from low-grade pulp to the most valuable hardwood for making furniture and veneer. Nor does the paper industry's appetite for wood mean, as some have suggested, that paper-making adds little value. In fact, paper leads the way in the Northern Forest in terms of economic activity generated by turning a tree into a finished product, and its jobs are among the highest paid blue-collar positions of any industry in the region. Moreover, the timber industry cannot survive solely by growing and cutting large-diameter trees that can take eighty years or more to mature. Landowners like Baskahegan and Wagner Woodlands – not to mention tree farmers like the Moffatts and logging contractors like Brian Sowers – must have markets for the smaller trees they thin out if they hope to stay solvent while larger trees grow to maturity.

Nonetheless, the paper industry's efforts to feed its mills and create steady quarterly profits exact a toll, in both the amount and the nature of the cutting these priorities inspire. The dominance of the paper industry in northern Maine and New Hampshire means that a large percentage of land in those states is devoted to a single goal: growing fiber as fast as possible. Niebling estimates that James River Corporation's pulp and paper mills in Berlin and Gorham, New Hampshire, for instance, consume the equivalent annual growth from one million acres of land. (James River itself owns fewer than seventy thousand acres in New Hampshire.) Some paper company lands are diverse forests from which pulpwood is thinned, but much of the acreage is managed on short rotations strictly for pulpwood. If applied too widely, such management degrades the forest, ignoring not only its broader economic potential but the biological and cultural richness a more diversely managed forest provides.

As the Northern Forest again enters a cycle of heavy use – a cycle that may determine its economic viability and ecological integrity for generations to come – we must do everything reasonable to encourage long-term forest management. The Northern Forest Lands Council's recommendations included some good suggestions, such as stabilizing property and estate taxes, establishing green certification programs, and increasing subsidies to landowners for protecting wildlife, rare plants, and other publicly valued natural resources. To the extent that these recommendations and other efforts encourage cultivation of the forest's long-term potential, they will strengthen the region's economic and ecological health. Such measures will also reduce the pressure for inappropriate development because, ideally, landowners will be able to earn enough from responsible forest management to retain their lands as working forests rather than carving them up for house lots.

The tougher question is how to halt the timber liquidations that compromise long-term forest health and often lead to land conversion. Two of the most commonly proposed means to stopping destructive forestry are the creation of stricter logging regulations and the purchase of large public reserves. Both have legitimate roles – and considerable limitations.

Regulation is a notoriously blunt instrument. If written carelessly, laws intended to control timber liquidators could put unreasonable restraints on responsible woods managers, eliminating the thin profits of small private landowners like the Moffatts and scaring off the kind of commercial investors on which companies like Wagner Woodlands depend. Well-intended regulations can produce unexpected side-effects. For instance, Maine's 250-acre-clearcut law reduced the size of clearcuts, but it may also be leading to increased high grading (taking all the good trees) and "checkerboards" of small clearcuts and intact forest, a pattern that fragments wildlife habitat. Banning clearcuts altogether, on the other hand, would rob foresters of a necessary tool. Small clearcuts can introduce diversity into single-species stands that have grown back on abandoned farmland, help create needed browse for wildlife, and enable foresters to salvage valuable wood from pestilence. Critics have also called for laws to control high grading. This is a worthy objective, but one virtually impossible to achieve through regulations; with thousands of harvests going on at any one time, states lack the personnel needed to enforce even the laws already on the books. And the fact is, a poorly executed partial cut can cause more erosion and habitat disturbance than a carefully planned small clearcut.

Large-scale forest practice regulations are difficult to write and administer effectively, and those who circumvent existing forestry laws will do the same with tougher ones. These limitations don't make regulation useless – and it is encouraging to see the Northern Forest states moving to strengthen their laws. But the shortcomings of the regulatory approach are numerous, and the public should not consider new laws to be a panacea.

The other most frequently suggested way to protect forests from abusive logging and subdivision – the purchase of land for public preserves – also has distinct but limited utility. We need permanent preserves in the Northern Forest to protect biodiversity, maintain an ecological baseline with which to compare the working forest, and provide outdoor recreation in a region where the most popular areas are suffering from overuse. To ensure the protection of representative ecosystem types, land known to contain habitat for rare or threatened plants and animals, or to be of other ecological or scenic significance, should be purchased from willing landowners and placed in the public trust. These purchases should include a limited number of large tracts – most in the 100,000-acre range, and perhaps one or two somewhat larger. A system of such preserves across the landscape, connected where possible by trail and wildlife corridors, would do much to maintain ecological integrity, promote low-impact tourism, and diversify local economies.

Public acquisition, however, has serious limitations as a way to "save the Northern Forest." The primary limitation is one of scope: Doubling the existing amount of public land in the twenty-six million-acre Northern Forest region – a frequently proposed target – would bring the total to only eight million acres, less than a third of the total area. However, buying four million additional acres would almost certainly require public investment of well over a billion dollars, and that kind of money – which would have to come from the federal government, because the states just don't have it – could be won only if environmental organizations committed all their political and financial capital to an immense national campaign. If the groups didn't demonstrate extraordinary sensitivity to the fears of local residents and the concerns of industry, they would alienate the very people whose cooperation they need to create the intricate mix of market incentives and public policies that would be necessary to conserve the remaining eighteen million acres. Any preserves that were created would thus become islands in an uncertain sea, and most of the Northern Forest would be left vulnerable to the same pressures that so many people are now trying to relieve.

The challenge, then, becomes not how to "save the Northern Forest," but how to protect these woods by using them well.

Unfortunately, this is a problem that Americans have not adequately solved, despite a century of conservation activism. Part of the problem is that neither of the environmental movement's two monumental accomplishments – the establishment of a system of public park, forest, and wilderness lands (mostly in the first half of the twentieth century) and the passage of some of the world's most stringent pollution laws (mostly in the century's second half) – addresses this fundamental question. Indeed, many of our public lands have been poorly managed and heavily exploited, notable examples being the overcutting of ancient forests in the west and in Alaska, and overgrazing and mining on Bureau of Land Management holdings. And while the Clean Air and Clean Water Acts curtailed the most egregious "point" sources of pollution such as smokestacks and discharge pipes, contamination by "nonpoint" sources such as poorly maintained automobiles and agricultural runoff continues largely unchecked. Thus while our public land system and pollution laws are substantial, even visionary, accomplishments, neither prepared the environmental movement to advance public policy that will help people and companies use land well.

The problem of how to use land well differs in fundamental ways from the issues environmentalists tackled in the great public lands and pollution campaigns. The strategies and tactics used to win those victories – every aspect, beginning with the way the problems were defined to the specific weapons used to fight the battles – sprung logically from the essential conflict in those issues, which was not merely a clash between resource use and the environment, but, more centrally, between the relatively narrow private gains of exploiters and the broader interests of the general populace. In the case of the early wave of public land acquisitions, extractive industries were decimating some of the nation's most spectacular landmarks; with pollution, industries were befouling the public's essential water and air.

It was the clarity and relative simplicity of these conflicts, both as the conflicts were portrayed and as they actually existed, that enabled environmentalists to enlist the support of the vast middle ground of citizens who might not otherwise pay attention to environmental problems, but who would rally to save clearly defined public resources from exploitation by narrow private interests. To generate this political support, environmentalists have usually employed compelling rhetoric, an appeal to a sense of urgency, and artful use of mass media. The drives to create the Adirondack Park and the national park and forest systems at the end of the nineteenth century are early examples of these tactics. A more recent campaign – one that served as a model for many to follow – was the Sierra Club's spectacularly successful effort in 1966 to stop a pair of proposed dams on the Colorado River that would have flooded extensive parts of the Grand Canyon. The club bought full-page advertisements in the *New York Times* and the *Washington Post* warning readers that "This time it's the Grand Canyon they want to flood" and that "Now only you can save Grand Canyon from being flooded . . . for profit." The ads inspired one of the biggest letter-writing efforts Congress had ever seen, and stopped the project. A similar mix – an appeal to aesthetics, the identification of a heinous evil being perpetrated by callous and powerful interests, and the communication of a sense of dire urgency – has since rallied public support in countless campaigns. If at times these crusades indulged in hyperbole, they nonetheless reflected the principal conflicts at hand.

When the forces driving land-use problems are more complicated, as they are in the Northern Forest – and particularly when much of the land in question has long been owned, lived on, and used by private citizens – efforts based on such simplistic dichotomies are bound to fail precisely because they ignore the essence of the situation, which is its almost overwhelming complexity. Appeals to "Save the Northern Forest" from a monolitic evil disregard the long, complex relationships between the land and the people who live here; the dedication of those who try to use the forest responsibly; and the vital links between residents and the land on which they live, the resource-based economies in which they work, and the communities in which they reside. These relationships are as complex, ambiguous, and confounding as nature itself. When environmental groups ignore this complexity and frame land-use issues in simplistic terms, they perform an act of reduction comparable to that practiced by landowners who consider the forest to be just so much lumber and pulp.

Such oversimplification is not lost on the people who live in these places. When environmentalists condemn entire extractive industries, impugning by implication those who make their living within these industries, they alienate the very people who could be their most passionate and effective allies. As we've talked to people in the Northern Forest, we've found many who share the basic goals of environmentalists, but who resent being cast as villains in a script that presents environmentalists as the saviors of land under siege. Both Chuck Gadzik of Maine and Leo Roberge of New Hampshire, for instance, are insulted at being identified as bad guys because they work in the forest industry or hunt, and that sense of insult has alienated them from the environmental community. Many loggers who try to do their jobs responsibly but who work under crushing economic pressures feel the same way. This is a shame, for it is among these people, who not only want but need the land to be treated with foresight and care, that environmentalists should find their best partners.

The resentments of those who work in resource-based industries are only sharpened by the environmental community's frequent use of aesthetic appeals to raise public support and funding. The description of breathtaking beauty, the photo of a pristine scene, the quote from Thoreau or Whitman or Muir – all meant to elicit a vision of nature as the realm of a poetic, epiphanous experience – have long been staples of the modern environmental campaign. To those who experience forests primarily in these ways, such appeals may set off reverberations. But to people who make their living in the woods, these invocations stand out less for their emotive power than for their incognizance of practical matters. By emphasizing the beauty of a place to the exclusion of almost all other qualities (except the oft-cited but poorly understood "biodiversity"), aesthetic appeals suggest that local residents lack the intelligence, education, and refinement to appreciate their own environment. Small wonder that people like Leo Roberge consider the "Sierra Club types" to be a privileged elite from elsewhere (or with income from elsewhere) with nothing more serious to worry about than where to watch birds on the weekend.

In alienating local people, albeit unintentionally (or, ironically, with the best of intentions), these campaigns lose the middle ground of public opinion. This middle ground is then aggressively sought and often won by free-enterprise and property-rights ideologues who apply their own economic and emotional leverage to convince locals that their interests don't mesh with those of the "tree-huggers." This is what happened in two notable land-use conflicts of the late 1980s and early 1990s: the "spotted owl" controversy over the old-growth forests of the Pacific Northwest, where environmentalist rhetoric alienated loggers and local businesses, who were then convinced by the large timber companies and Reagan-Bush administrations (who had created the problem in the first place) that those arguing for forest protection valued owls more than people; and in the Adirondacks, where the Cuomo Commission's insensitivity to local attitudes made it easy for developers and private-property-rights activists to convince many residents that the environmentalists were elitists concerned only with preserving their outdoor playground.

Similar confrontations – and resulting misunderstandings – have occurred on a smaller scale in the Northern Forest.

In January 1989, less than a year after the Diamond land sale, the Wilderness Society called for the establishment of a 2.7-million-acre public reserve in Maine. On its surface, the Maine Woods Reserve plan made sense. Maine had the least amount of public land in New England, just 5 percent of the land base. The proposed reserve would be centered around the state's largest existing public holding, 200,000-acre Baxter State Park, and would encompass the watersheds of several important rivers, much of the shoreline of massive Moosehead Lake, and parts of the Appalachian Trail. All of these areas have outstanding natural and recreational features. Much of the land was owned by large commercial timber companies, some of which were considering selling at the time the Wilderness Society plan came out. Maine had always been an environmentally conscientious state, and the preserve idea was sure to appeal to residents in the more urbanized parts of southern Maine. The problem was, the Wilderness Society announced its plans to the press without meeting beforehand with the major landowners, public officials, or area residents. At the time the group had no office in Maine, so its press releases carried Boston and Washington, D.C. addresses. A regional office in Boston had opened in 1985, but the staff had spent little time in Maine. So, though the proposal had genuine merit, the group hadn't done the groundwork necessary to gauge or influence local opinion. For instance, the Society badly underestimated the property-rights movement, which had been stirred a year earlier by another national group's plan to preserve parts of Maine's scenic northern coast. As a result, opposition to the Society's proposal was swift and rigid; not even Maine's statewide conservation groups, which also had been caught off guard by the announcement, publicly embraced the plan. While the Society's plan helped galvanize some sectors of the public, the group simply could not counter the impression among residents of central and northern Maine that a national organization was trying to dictate what was best for them. Neither could it combat the many simplistic distortions of its proposal. The Society recognized these problems within a matter of months and hired a Maine native named Jym St. Pierre, a forest expert with a reputation for frankness, to open an office in Augusta. But the damage had been done. For years afterward, the Wilderness Society was synonymous with "meddlesome outsider" in the minds of many Mainers. It didn't help matters when, in 1993, the Society closed the Maine office.

A much smaller group, Restore: The North Woods, also operating out of Massachusetts, managed to infuriate not just industry but much of the conservation community in 1994 with the high-handed way it filed a petition to list the Atlantic salmon as a federally listed endangered species. As a result of damming and pollution, the salmon's U.S. populations had dropped since pre-Colonial times from hundreds of thousands to around three or four thousand. Virtually everyone agreed that *Salmo salar,* which is born in rivers, swims to the sea to mature, and then returns to inland rivers to spawn, was in trouble. In fact, fishing and conservation groups had been working throughout New England since the 1940s to bring the salmon back, and had created a cooperative network of advocates and programs. While these efforts had not restored the species to fully viable populations, the network had rescued the salmon from the edge of extinction in the 1960s and was a productive partnership between anglers and conservationists.

When Restore decided to file its petition with the U.S. Fish and Wildlife Service, however, the group alternately ignored and insulted this coalition. Restore staffer David Carle tapped some of the salmon groups for information, but he told few of them of Restore's plans to seek an endangered listing. If Carle had asked, most of the salmon groups would have told him that they considered classifying the fish as endangered a terrible idea (a reaction he perhaps anticipated). Many of the organizations said privately that if consulted beforehand, they would have supported a "threatened" listing, which would have given the fish most of the same protections as endangered status, but put fewer restrictions on relatively benign activities such as catch-and-release sport fishing. But the endangered species petition fell on a majority of the fishing and conservation groups as a complete surprise. Some of Carle's subsequent public statements further enraged and embarrassed the conservationists who had worked so long on the salmon's behalf. Noted one endangered-species specialist, "The petition has put the lobbyists for the salmon, the fishermen, into the same camp as the paper and power companies, because they are now fighting a common enemy here: the petition." Restore's handling of the petition drove away some of the group's best potential allies. It also irked many environmentalists involved in the larger Northern Forest debate, who worried that an endangered listing would lead to a "salmon versus jobs" battle that could destroy attempts to get environmentalists, industry, and locals to work together.

Even groups that have established good relations with locals can find their efforts undermined if they fail to take each issue straight to the people who live in the environment in question. In 1993, the Appalachian Mountain Club, a Boston-based recreation and conservation group with an extensive hut and trail system in New Hampshire's White Mountains, joined a coalition of eight other organizations that were intervening in the Federal Energy Regulatory Commission's (FERC) relicensing of sixty-four hydropower dams in

New England. The dams up for review included seven on a stretch of the Androscoggin River in Berlin and Gorham, New Hampshire – five owned by James River Corporation and two by Public Service Company, the state's largest electric utility. The dams sometimes choked river flows to a trickle, and residents of one riverside neighborhood in Gorham contended that the exposed river bed was releasing dioxin (a common byproduct of papermaking) that was contributing to elevated cancer rates. The AMC and the other intervenors wanted the new FERC license to require adequate flow for fish and canoeing, protected shorelines, energy-conservation equipment, fish ladders, and improved public access to the water.

Despite making a strong alliance with residents of the affected neighborhood, however, the AMC was immediately attacked for its efforts by several local officials and private-property-rights activists. Led by a former AMC employee and Gorham selectman named Michael Waddell (who was aided by a sympathetic local newspaper), the AMC's critics insisted at every opportunity that the environmentalists were using the FERC process to undermine the forest products industry and turn the Northern Forest into a park for hikers from Boston. Distracted by other priorities, the AMC did not act quickly or forcefully enough to rebut this unfair charge. For instance, no AMC staffer attended an early FERC hearing in Berlin at which Waddell characterized the organization as elitist and out of touch with local concerns. When the club realized the extent of the problem, its senior staff tried to remedy the situation by writing letters to local newspapers, hosting a mini-conference on the Northern Forest for area residents, arranging a series of facilitated meetings with local officials, and helping to plan a Northern Forest Heritage Park in Berlin to celebrate the area's history. The club dropped its complaint in New Hampshire after an initial ruling by FERC in the summer of 1994 indicated that some of the requested mitigation would come through. Yet by that fall, a few local officials were still giving the AMC trouble by threatening to intervene in the

U.S. Forest Service's repermitting of the club's White Mountain hut system – despite that the huts employ more than a hundred workers during the summer and attract tens of thousands of tourists who support the area's economy. So while the AMC had been an important economic and social presence in the area for more than a century, their failure to make their case adequately to residents during the critical stages of the FERC controversy left them open to their enemies' distortions.

The environmental community engages in much hand wringing about its political credibility in the face of such debacles, and properly blames property-rights ideologues and avaricious special interests for inflaming fears and encouraging misunderstanding. But the groups do too little to prevent such problems in the first place. Environmentalists should consider the alienation of local interests to be a major liability not just because it causes discrete, localized political crises, but because alienation of local people signals that conservationists may be losing the middle ground to the far right. Conservationists must acknowledge that tangible support for their cause is soft: Three-quarters of Americans tell pollsters they are environmentally concerned, yet fewer than 5 percent donate money or time to environmental groups, and ecological issues rank low in perceived importance to people's daily lives. If environmental groups don't want to cede the precious and vulnerable middle ground – and they should not – then they should devote as much financial and political capital to forming alliances with local residents, responsible business interests, and key community leaders as they do to courting new members and donors.

One of the most effective ways environmentalists can create such partnerships is to focus on initiatives that combine good environmental policies with local economic and public health interests. Some groups have already done so in fighting polluters, combining their legal, political, and scientific expertise with the knowledge, determination, and authenticity of residents who suffer the consequences of air and water contamination. Environmentalists must do likewise, even more assertively and on a sustained basis, in land-use issues.

Forming these alliances will require recognizing that, important though they are, abstract values such as biological diversity mean little to a logger struggling to survive in an economic system that favors production over everything else. Even so, loggers face problems that environmentalists could help solve and, in so doing, advance their own causes. For instance, the high cost of workers' compensation insurance is not merely a business problem for logging contractors; by reducing the contractors' small profit margins, these costs often pressure loggers to cut faster and more carelessly than the loggers would wish. An environmental group could create an effective alliance with loggers by working with them to win workers' compensation reforms that substantially decrease rates while rewarding safer logging practices. By doing so environmentalists would show, rather than merely preach, the interdependence of sensible economic and environmental policies; remove a significant disincentive to environmentally responsible logging; gain a greater understanding of the gritty pocketbook issues that worry local people; and win friends who might be helpful when more contentious issues arise. Similar opportunities exist elsewhere: in cooperatively developing green certification programs for wood products; in convincing lending institutions to offer lower-interest, longer payback terms to timberland investors; in helping to produce educational programs for foresters and loggers, which can impart useful safety, business, and ecological knowledge; in forming partnerships to encourage and support new value-added industries, which can create new jobs and help provide the more diverse forest economy needed to encourage more diverse forest management.

Conservation by mass-media campaign, though it might bring attention and win specific battles, will not win these crucial alliances. What is required is to go into the communities and woods and work with the people there, learning about the resource-based economy and helping to nurture its environmentally beneficial aspects while discouraging the more destructive elements. This will mean having the courage to make some mistakes. It will mean entering arenas in which environmentalists must learn as well as teach. Most important, it will mean letting locals have at least an equal hand in shaping the agenda.

Local residents must bear responsibilities as well. In the words of environmental activist Jamie Sayen, who lives in the heart of the Northern Forest in New Hampshire and has had his share of disputes with neighbors, "Locals are stupid to say that the greens are the problem, and the greens are stupid if they think they are in touch with the needs of people up here. I think the locals have been screwed over by big timber interests, but to a certain extent they are guilty of allowing this to be done to them. And the environmental community has made the worst of a bad situation. We have taken natural allies and made them hostile. Sometimes you have to deliver bad news to folks up here and say things they don't want to hear. But if you are honest and firm, they will respect it. Disagreement is healthy as long as we're disagreeing over the real issues."

We don't mean to suggest that everything can be solved solely by working more closely with local interests. Many problems in the Northern Forest are caused by complex global trends and can only be addressed by new policies in the state capitals, in Washington, and on Wall Street. But to get such initiatives passed, environmentalists will need local allies. Particularly in lightly populated states such as those in northern New England, a few determined residents can be a very effective lobby with a governor or member of Congress. Until conservationists demonstrate that they recognize that top-down approaches must complement rather than supercede locally driven solutions, many local residents will continue to resist policies imposed from the outside, even if those policies ultimately make sense for their lives.

If environmentalists are to build lasting alliances with local people they must also persuade two of their most important existing allies – donors and the media – to respond to more sophisticated, carefully considered appeals. Donors must be shown that serious environmental problems are not necessarily battles between good and evil. And the press – whose attention environmental groups can usually attract only by using rhetoric so dramatic that it opens the groups to charges of being alarmist – must be convinced to present environmental problems in ways that don't distort the subtlety of the issues involved for the sake of good copy.

The changes we're suggesting will require a major philosophical adjustment in the environmental community, for granting primacy to local influence, ideas, and interests contradicts the conservation movement's fundamental belief that resources are best conserved through scientifically informed, centralized decision making. As noted by Samuel P. Hays in his seminal history, *Conservation and the Gospel of Efficiency,* it was this belief in centralized solutions that drove the major early successes of the American environmental movement, such as the Weeks Act (which first established the eastern National Forest system). This belief has subsequently inspired most of the movement's substantial accomplishments, including federal initiatives such as the National Environmental Policy Act (which guides federal land management), comprehensive pollution laws such as the Clean Water Act and the Clean Air Act, the Endangered Species Act, and the establishment of the Environmental Protection Agency.

These laws and programs have done much to protect the environment, and they must be maintained and strengthened. In this age of the finite, however, we may be reaching the limits not just of what our lands can bear, but of the strategies and tools we have long used to protect them.

We stand, really, at the beginning of another epoch of environmentalism – what one might call the movement's third era. Activists who led the first two eras are rightly credited with the major successes described earlier: in the first half of this century, the establishment of public lands and the emergence of scientific resource management; in the second half (especially in the 1970s and 1980s), vital legislative and regulatory victories.

As this new era begins, land-use and other environmental issues are more complex than ever before, involving increasingly complicated conflicts between private and public interests, particularly in the use of resources such as forests and fisheries. To solve these problems, we must develop new solutions in the same way that good foresters manage woodlands: from the ground up, and in response to local conditions.

This will mean decentralizing many decisions by truly involving local people, especially those who work in resource-extractive industries. We need to do this for more than just political reasons. We need to do this because those who actually try to manage these resources – tree farmers, contract loggers, foresters, wildlife managers and biologists, commercial fishers, and the like – can contribute uniquely valuable experience, knowledge, and perspective. For those are the people who struggle constantly with the question that has always lain at the core of environmentalism: How do we balance use and conservation of our resources?

For most of us, this question of balance takes abstract intellectual, emotional, or political forms. Those who work the resources, however – who are called upon to pull from the earth what the rest of us demand – face this question all the time. Some ignore the question and never attempt to strike a sensible balance. But many, including the loggers, foresters, and landowners in this book, do try. They strive every day to answer this question responsibly, even when at times to do so compromises their own situations and interests.

This new and difficult work will require tremendous patience. The job of getting to know and working with residents of areas such as the Northern Forest – of trusting them, of earning and deserving their respect – is not easy or comfortable. But if environmentalists are to ask foresters, loggers, and landowners to be patient with the forest, they should ask nothing less of themselves when dealing with the forest's human inhabitants.

Already, some in the environmental community recognize this. The Vermont Natural Resources Council has spent considerable portions of its limited resources meeting with citizens in the Northern Forest, not just to gain their support, but to understand and help address their concerns. The VNRC also commissioned and published a fine, unbiased study of one area's forest-based economy that reveals much about the pressures on landowners, loggers, and wood products manufacturers, and makes clear the vital role that value-added manufacturing plays in promoting good forestry. The Adirondack Council, despite being under constant fire from property-rights groups, has worked with Adirondack towns to help secure funding for such community needs as water and sewage treatment and landfill improvements. After the Northern Forest Lands Council's report was completed, the Society for the Protection of New Hampshire Forests created a coalition with the New Hampshire Timberland Owners Association and the North Country Council (an economic development and planning agency formed by northern New Hampshire towns) to jointly educate the area's residents and advance the NFLC's recommendations. The Maine Audubon Society joined forces with hunting and fishing clubs on a model initiative to promote land conservation through proceeds of the state lottery. Groups in each state are also encouraging local land trusts to identify and protect special areas through community efforts. Restore: The North Woods, which had run into trouble with its 1993 salmon petiton, reached out to local people in building support for its 1994 proposal for a large national park in Maine. And many environmental groups tried to assess local perspectives by carefully reviewing and compiling comments made by local residents about the Northern Forest Lands Council's report.

Public agencies are also recognizing the importance of meeting early and often with residents of areas proposed for conservation. One reason Leo Roberge and many of his friends came to support the Umbagog National Wildlife Refuge is that Dick Dyer of the U.S. Fish and Wildlife Service spent countless evenings in the North Country explaining the plan. Federal officials in charge of the proposed Silvio Conte National Wildlife Refuge along the 400-mile Connecticut River are doing the same. In 1993 and 1994, the U.S. Fish and Wildlife Service held an astonishing 160 meetings in towns from the river's source in Pittsburg, New Hampshire, to its outlet at Long Island Sound. These meetings were designed not to present a preconceived plan, but to solicit ideas on how the refuge, which was proposed by an act of Congress, should be put together. Despite rigid objections by some, it became clear during these meetings that most people were in favor of some sort of protection. Even those who felt the refuge was not necessary were mollified because they were given a chance to voice their objection early in the process, and because it was their neighbors – and not bureaucrats or environmentalists "from away" – who were developing the plan. Project leader Larry Bandolin calls such an attitude "informed consent." He says, "Some people were saying that they didn't like the idea, but they were going to go along with it because they had been listened to."

Handing decision-making to local residents can be extremely frustrating, and there are bound to be failures because some people will not participate in good faith. Attempts to designate several New England waterways as national Wild and Scenic Rivers in 1993 and 1994 were defeated by campaigns of distortion and intimidation by fringe property-rights groups in New Hampshire, Maine, and Massachusetts. But such setbacks should not be taken as excuses to abandon the long-term goals of building local alliances. Indeed, a key official from the National Park Service who was intimately involved with the Wild and Scenic nominations in New England says that despite the failure to establish federal protection, the local committees set up to study the issue ended up writing far-reaching conservation plans that are now being put to use.

"We have to be open-minded in terms of what the solutions are, while being absolutely committed to change," says Stephen Blackmer of the Appalachian Mountain Club, who says he learned valuable lessons from his organization's problems in northern New Hampshire. "If we're not open-minded, if we insist it has to be done this way and not that way, we're lost. We have to stick our necks out. But we have to stick our necks out with humility. That's a hard thing to do."

Perhaps most encouraging, an awareness of the need to work patiently with local people and private landowners has spread to the bioreserve or "wildlands" movement – the growing, worldwide push to establish an international system of wild reserves, connected by protected corridors, to help preserve species diversity and biotic integrity. This is one of the most ambitious environmental "campaigns" (the word scarcely applies to such a long-reaching goal) ever undertaken. It is driven primarily by an alliance between conservation biologists, headed by Michael Soulé, and what might be considered "radical" preservationists, most notably Dave Foreman, the former leader of Earth First! The wildlands concept has met stubborn resistance from many hard-line industry people and property-rights advocates who see (or claim to see) the movement as a giant land grab carried out by ideologues out of touch with political reality. Fortunately, while some in the wildlands movement do seem lost in the righteousness and apparent urgency of their mission, others, including Soulé, clearly see the need to move slowly and with full recognition of local people's fears. In a 1993 special "Wildlands Project" issue of the environmental journal *Wild Earth,* Soulé explains why he would speak in terms of fifty- or one hundred-year periods when trying to convince a forestland owner or a National Forest supervisor to change their practices and gradually give up appropriate land to bioreserves:

> Some [environmentalists] will ask why we should adopt . . . a policy of patience. The answer is fear – fear on the part of those folks who be-

lieve they will lose their jobs as loggers or miners, have to abandon their way of life as ranchers, professional guides or commercial fishermen, and be forced to move from the region where their families have been living for generations. As we all have learned, fear translates into potent political opposition.

Maybe our task as conservationists is to remove the fear from people who see themselves threatened by attacks on their occupations, their livelihood, their world view and their property. The first step is to admit that much of their fear is born of impatience – *our* impatience. After all, 100 years is less than $1/10,000$ of the lifetime of the average vertebrate species. The goal should be staying the course, not setting a speed record.

The patience Soulé advocates will be essential to building new alliances – alliances that combine the influence that organized environmental groups can wield in centers of political power, with the wisdom and support of people who draw their sustenance from the land every day.

The Northern Forest, with its wide, deep history, its tradition of face-to-face democratic decision-making, its resilient woods and resourceful people, offers a singularly promising place to start. Here in America's oldest working woodlands we can try to renew not just the land but environmentalism itself – and in doing so, restore not just these woods, but our relationship to them and to one another.

GLOSSARY

BASAL AREA: A method of measuring and expressing the density of trees per acre in a forest. Basal area is the combined horizontal surface area of the trees (meaning the tops of the stumps if all the trees were cut down) at 4' 6" off the ground.

BIOLOGICAL DIVERSITY, OR BIODIVERSITY: There have been many attempts to define this concept. Among the most useful explanations we have found are one adopted by the Society of American Foresters (SAF), and another offered by Dr. Stephen Trombulak, a noted conservation biologist at Middlebury College in Vermont. The SAF defines biodiversity as "The variety and abundance of species, their genetic composition, and the communities, ecosystems, and landscapes in which they occur. It also refers to ecological structures, functions, and processes at all of these levels. Biological diversity occurs at spatial scales that range from local through regional to global." Trombulak argues that "Promoting biological integrity is more than just protecting a few high-profile endangered species or maximizing the number of species on every acre of land. It involves promoting the existence of all native species and ecosystems in their natural patterns of abundance and distribution, as well as the ecological and evolutionary processes that connect them over the landscape and over time."

BOARD FOOT: The common measurement for sawn logs in the United States. One board foot is a piece of wood one foot square and one inch thick.

CHAIN: A measurement for linear distance on the ground, used by loggers and foresters. A chain is equal to 66 feet.

CLEARCUT: A generic term used to describe a timber harvest in which all the trees are removed.

CONSERVATION EASEMENT: An increasingly popular land conservation technique that leaves land undeveloped while allowing certain specified uses. A conservation easement is a permanent, legally binding agreement between a landowner and a private or public conservation agency. By selling or donating an easement, the landowner relinquishes certain rights of ownership – generally the rights to subdivide, develop, or mine the property – but retains title to the land itself. The agency that receives the easement accepts the responsibility

344

to monitor and enforce the terms in perpetuity. Land under a conservation easement may be sold or otherwise transferred to other owners, but the restrictions on the land's use remain intact. In the Northern Forest states, easements usually allow forestry and agriculture.

ECOSYSTEM MANAGEMENT: An emerging approach to land management that instead of emphasing isolated objectives (such as producing specific volumes of timber or saving a certain plant or animal species), attempts to enhance the complex biological, hydrological, and ecological relationships and processes that contribute to overall forest health. Again, the Society of American Foresters' definition is among the most useful. It reads, in part, "[Ecosystem management] attempts to maintain the complex processes, pathways and interdependencies of forest ecosystems and keep them functioning well over long periods of time, in order to provide resilience to short-term stress and adaptation to long-term change. Thus, the condition of the forest landscape is the dominant focus, and the sustained yield of products and services is provided within this context."

FELLER-BUNCHER: A class of whole-tree harvesting machines which can grasp whole trees, cut through their trunks, and lay them down. The feller-buncher, which often resembles a backhoe, is run by an operator sitting in an enclosed cab. The mechanical arm that cuts the trees is equipped either with a circular blade or hydraulic shears.

HIGH GRADING: A generic term used to describe a timber harvest in which all the marketable trees are cut.

INDUSTRIAL LANDOWNER: A commercial landowner that also owns a wood manufacturing facility or facilities, such as a sawmill, pulp mill, or paper mill.

NON-INDUSTRIAL LANDOWNER: A landowner that does not own manufacturing facilities. While the term applies to any such landowner, including owners of small unmanaged woodlots, it is used here to denote landowners who sell wood to industry.

PULP: Wood fiber that is broken down by mechanical and/or chemical means and used in the manufacture of paper products.

PULPWOOD: Wood that is harvested and sold to make pulp. Pulpwood is transported from the forest to the mill as long logs, 4-foot logs, or chips. Depending on the manufacturer, pulpwood may be softwood or hardwood species, and can come in virtually any size. However, pulpwood generally consists of smaller, less well-formed logs that are unsuitable for sawlogs.

REGENERATION: The successive growth of seedlings, stump sprouts, or root suckers that ensues either as a stand of trees matures and propagates, or is removed or thinned by harvest, wind, fire, disease, pestilence, or natural mortality. The word "regeneration" is used to denote both the process and the vegetation that results.

RELEASE CUT: A method of timber harvest by which most or all of the mature trees are removed to increase light to the younger trees and seedlings, thereby "releasing" them from the repressing shade.

ROUNDWOOD: Any wood product that is transported from the forest to the mill as logs rather than as chips.

SAWLOGS: Sections of trees, virtually always the trunks, that are sold to be cut into lumber. Sawlogs are distinguished from logs that will be chipped or otherwise processed for pulp, fuel, or manufactured products (such as chipboard, particle board, oriented strand board, plywood, or veneer).

SELECTIVE CUT: A generic term used to describe any number of timber harvesting methods whereby individual trees are "selected" to be cut. Laypersons often use this term erroneously to denote a preferable alternative to other harvesting methods, especially "clearcutting" and "high grading" when, in fact, high grading is a form of selective cutting.

SHELTERWOOD CUT: A method of forest management by which the overstory – the trees with the tallest crowns – is removed in successive harvests, slowly increasing light to the trees below. It is most often practiced in forests of shade-tolerant species such as spruce-fir.

SILVICULTURE: The cultivation of trees in a forest.

SKIDDER: A machine used to haul trees and logs out of the forest after they have been cut down. Modern skidders ride on four large rubber tires and look somewhat like a cross between a bulldozer and a tractor. They are usually hinged in the middle to facilitate maneuvering between trees. Skidders gather and drag logs, trees, and piles of branches with either a cable and winch or a hydraulic grapple.

SUGARBUSH: A stand of sugar maple trees *(Acer saccharum)* that is managed for maple sugar and maple syrup by tapping the trees for sap in the early spring. The words *sugar* and *sugaring* are often used as verbs to describe the job of tapping the trees, collecting the sap, and making syrup.

SUSTAINED YIELD: A balance between harvest rates, growth, and natural mortality that enables a forest to produce timber indefinitely.

VALUE-ADDED INDUSTRY: Industries that "add value" to raw wood products by fashioning them into either materials such as lumber, or finished products such as furniture, paper, or other wood products.

Adding value to wood in this way – as opposed to exporting the raw logs – increases the local economic value of every tree cut. Generally, the manufacturing jobs that add the most value, such as furniture making, also create more skilled, better-paying jobs.

WHOLE-TREE HARVESTER: A class of machines that can fell trees. Whole-tree harvesting crews thus fell trees with machinery, rather than chainsaws. Some whole-tree harvesters (such as the Valmet Single Grip Processors used on the Baskahegan Company's land in Maine) also de-limb and saw the trunks into desired log lengths.

SELECT BIBLIOGRAPHY

Aplet, Gregory, Nels Johnson, Jeffrey T. Olson, and V. Alaric Sample, eds. *Defining Sustainable Forestry.* Washington, D.C.: Island Press, The Wilderness Society, 1993.

Beattie, Mollie, Charles Thompson, and Lynn Levine. *Working with Your Woodland: A Landowner's Guide,* Rev. ed. Hanover, N.H.: University Press of New England, 1993.

Brown, William Robinson. *Our Forest Heritage: A History of Forestry and Recreation in New Hampshire.* Concord, N.H.: New Hampshire Historical Society, 1958.

Bryan, Frank, and John McClaughry. *The Vermont Papers.* White River Junction, Vt.: Chelsea Green Publishing Co., 1989.

Connor, Jack. "Empty Skies." *Harrowsmith* 16 (July/August 1988): 34–45.

Coolidge, Philip T. *History of the Maine Woods.* Bangor: Furbush-Roberts Printing Company, Inc., 1963.

Cronon, William. *Changes in the Land: Indians, Colonists, and the Ecology of New England.* New York, N.Y.: Hill and Wang, 1983.

D'Elia, Anthony N. *The Adirondack Rebellion.* Onchiota, N.Y.: Onchiota Books, 1979.

Forest Information Center, Maine Forest Service, Department of Conservation, State of Maine. *1992 Silvicultural Activities Report.* Augusta.

Forest Information Center, Maine Forest Service, Department of Conservation, State of Maine. *1991 Silvicultural Activities Report.* Augusta.

Graham, Frank, Jr. *The Adirondack Park: A Political History.* New York: Alfred Knopf, 1978.

Harper, S.C., L.L. Falk, and E.W. Rankin. *The Northern Forest Lands Study of New England and New York.* Rutland, Vt.: U.S.D.A. Forest Service, 1990.

Hays, Samuel P. *Conservation and the Gospel of Efficiency.* New York: Atheneum, 1979.

Irland, Lloyd. *Wildlands and Woodlots: The Story of New England's Forests.* Hanover, N.H.: University Press of New England, 1982.

Irland, Lloyd. *Wildlands and Woodlots: The Story of New England's Forests.* Rev. ed. Unpublished manuscript, 1993.

SELECT BIBLIOGRAPHY

Johnson, Charles W. *The Nature of Vermont.* Hanover, N.H.: University Press of New England, 1980.

Klyza, Christopher McGrory, and Stephen C. Trombulak, eds. *The Future of the Northern Forest.* Hanover, N.H.: University Press of New England, 1994.

Kricher, John C., and Gordon Morrison. *Eastern Forests.* The Peterson Field Guide Series. Boston: Houghton Mifflin Co., 1988.

Lansky, Mitch. *Beyond the Beauty Strip: Saving What's Left of Our Forests.* Gardiner, Me.: Tilbury House, 1992.

Manning, Richard. *Last Stand.* New York: Penguin Books, 1991.

Marchand, Peter. *The North Woods: An Inside Look at the Nature of Forests in the Northeast.* Boston: Appalachian Mountain Club, 1987.

Milliken, Roger, Jr. *Forest for the Trees: A History of the Baskahegan Company.* Cumberland, Me.: Roger Milliken, Jr., 1983.

Nash, Roderick. *Wilderness and the American Mind.* Rev. ed. New Haven.: Yale University Press, 1973.

Northeastern Forest Alliance. *The Economic Importance of the Northeast Forest.* Saranac Lake, N.Y.: Northeastern Forest Alliance, [1992].

Northern Forest Lands Council. *Finding Common Ground: The Draft Recommendations of the Northern Forest Lands Council.* Concord, N.H.: Northern Forest Lands Council, March 1994.

Northern Forest Lands Council. *Finding Common Ground: The Recommendations of the Northern Forest Lands Council.* Concord, N.H.: Northern Forest Lands Council, September 1994.

Northern Forest Lands Council. *Technical Appendix: A Compendium of Technical Research and Forum Proceedings from the Northern Forest Lands Council.* Concord, N.H.: Northern Forest Lands Council, 1994.

Ober, Richard, ed. *At What Cost? Shaping the Land We Call New Hampshire.* Concord, N.H.: New Hampshire Historical Society, Society for the Protection of New Hampshire Forests, 1992.

Sayen, Jamie, ed. *The Northern Forest Forum,* Volumes 1, 2, 3. Lancaster, N.H.: Northern Appalachian Restoration Project. 1992–1994.

Sherman, Joe. *Fast Lane on a Dirt Road: Vermont Transformed, 1945–1990.* Woodstock, Vt.: Countryman Press, 1991.

Smith, David C. "Continuity Lost." *Habitat: Journal of the Maine Audubon Society* Volume 7, Number 3 (June 1990): 24–26.

Terborgh, John. "Why American Songbirds Are Vanishing." *Scientific American,* May 1991: 98–103.

U.S. Department of Agriculture. U.S.D.A. Forest Service. *Forest Resources of the United States, 1992.* Fort Collins, Co., 1993.

INDEX

Act 200, Vermont, 198
Act 250, Vermont, 199, 207-10, 261
Adirondack Council, 286, 292, 296, 306, 339
Adirondack Daily Enterprise, 287, 289
Adirondack Defender, 281
Adirondack Forest Preserve, 275, 277
Adirondack Liberators, 292
Adirondack Minutemen, 281
Adirondack Mountain Club, 268
Adirondack Park, 272-98, 328
Adirondack Park, The (Graham), 277
Adirondack Park Agency (APA) Act, 277-79, 281, 285-86
Adirondack Park Agency, creation of, 277-79; political disputes between Adirondack residents and, 272
Adirondack property rights movement, 279-98, 331
Adirondack Rebellion, The (D'Elia), 280
Adirondack Solidarity Alliance, 271-72, 291
Allen, Dave, 218-20, 223-24
Alliance for America, 271
Amati, Nicholas, 247
American Forest Council, 225, 227, 228
Androscoggin Fish and Game Club, 55
Androscoggin River, 3-4, 6, 8, 48-49, 56; as habitat for songbirds, 36-40; relicensing of dams on, 334-35; as route for log drives, 19, 65
APA Act, *see* Adirondack Park Agency Act
Appalachian Mountain Club, 306, 308, 333-35, 341
Association of Adirondack Towns and Villages (AATV), 289
Atlantic salmon, petition to list as endangered species, 157, 307-308, 332-33

Audubon Society of Maine, 120, 306, 339; distances itself from Restore: The North Woods, 309
Audubon Society of New Hampshire, 31, 33, 36, 41, 306
Aziscoos Lake, 3

Bandolin, Larry, 340
Bangor, Maine, as early logging center, 63-64
Barre, Vermont, 176-77, 230
Bartlett, Colin, 87-88
Baskahegan Company, 110, 118, 128, 133, 142, 322-24; contracts with Treeline, Inc., 96; history of, 67-72, 79-83; management of lands under Chuck Gadzik, 73-78, 83-92, 147, 149-59
Baskahegan Dam Company, 67-69; *see also* Baskahegan Company
Baskahegan Lake, early logging around, 62-65, 67
Baskahegan Stream, 61-63, 67
Bates College, 155-56
Baxter State Park, 87, 155
Bendick, Robert, 268-69
Berle, Peter, 286-87
Berlin, New Hampshire, 4, 7, 15-25
Bevin's Marine and Cycle Center, 230-31
Beyond the Beauty Strip (Lansky), 117
Bingham, William, 64
Blackmer, Steve, 341
Bley, Jerry, 117, 125, 128, 132
Blue Line, Adirondack Park, 275, 277, 290
Bofinger, Paul, 315-18
Boise-Cascade, 21, 28, 96
Bon Appetit, 203-204
Boston Globe, 203-204

350

Index

Bradford, William, *xvi*

Brattleboro, Vermont, as major sawlog center, 179

Brown Company, *xviii*, 15-21, 44, 300

Brown, Larry, 244-45, 322

Brown, William Robinson (W.R.), 17, 18

Brown, William Wentworth, 19

Bryce, Phil, 28

Budbill, David, 259-60

Buntline, Ned, 273

Burlington Electric Department, 240

Burlington, Vermont, as major sawlog port, 179

Butternut Country Store, 226

Cambridge River, 3

Canada, large tree farms in, 214-15; maple syrup industry in, 229

Carbonetti, Richard, 303-304

Carey, Governor Hugh, 282

Carle, David, 157-58, 310, 333; *see also* Restore: The North Woods

Carlisle (family): ownership in Maine, 65

Carlisle, George, 70

Carpenter, Sterling, 206-207

Carter, President Jimmy, 82

Champion International Corporation, forest management, 122-24, 125, 142; contracts with Baskahegan Company, 74-76; participation in Project SHARE, 157

Chandler, Bob, 144

China, as buyer of New England wood, 320

Chute, Carolyn, 259; *Merry Men*, 250

Cie Générale Electricité (CGE), *xx-xxi*

Citizens Group of the Adirondacks, 288-89, 291

Civil War, 66

Clean Air Act, 327, 338

Clean Water Act, 327, 338

Cline, Mike, 120, 126-28, 131-33

Coburn (family): ownership in Maine, 65

Codling, Dee, 231

Coe, Benjamin, 267, 272

Colvin, Verplanck, 273

Commission on the Adirondacks in the Twenty-First Century, *see* Cuomo Commission

Concord Woodworking Company, 239

Connecticut River, *xv*, 179, 340

Conservation and the Gospel of Efficiency (Hays), 338

Coos County, 3, 20

Cosmopolitan, paper for, 75

Cote, Luke, 29

Cowles, Esther, 301-302

Craftsbury Academy, 194-95

Craftsbury Horse and Buggy Club, 199-200

Craftsbury Old-Time Fiddlers Contest, 199-202

Craftsbury Sports Center, 199, 205-206

Cuomo Commission, 285-93, 301, 331

Cuomo, Governor Mario, 287, 294, 296-300

Cureton, John, 135

Cushman, Paul, 77, 84

D'Elia, Anthony, 280-81, *The Adirondack Rebellion*, 280

Davis, George, 286

Davis, Pat, 237-38

De Geus, Bob, 190

Dead River, as route for log drives, 65

Dead River Company, 72, 79-82, 84, 152

Dean, Edgar, 21

Dean, Governor Howard, 304

Declaration of Independence, 262

Demars, Larry, 216-18, 220-24, 232

Diamond International Corporation, *xv*; *see also* Diamond land sale

Diamond land sale, *xx-xxiii*, 24, 27, 285, 300, 319, 321, 331

Diamond River, 3, 18

Difley, Jane, 227-28

Dube, Mike, 101-104

Ducks Unlimited, 51

Dyer, Richard "Dick," 29, 340

Eames, Everett, 29

Earth First!, 292

East Grand Lake, 143

Eastern Manufacturing Corporation, 67-69

Eden, Vermont, 175, 177

Emery, Bill, 108-109

Endangered Species Act, 157, 332-33, 338

Environmental Protection Agency, 338

Erie Canal, 274

Errol dam, 3-4

Errol, New Hampshire, 3, 27

Index

Esden, Jim, 193
Ethan Allen Furniture Company, 227

Fair, Jeff, 29, 30-35
Fair, Tripper, 30
Federal Energy Regulatory Commission
  hydropower licenses, 24, 33, 157, 333-35
Finch-Pruyn, 291
Findings and Options (Northern Forest
  Lands Council), 302
Finding Common Ground (Northern
  Forest Lands Council), 267, 300, 310-18
Fitzgerald, Charles, 117
Flack, Robert, 291
Flood, Pat, 124-25, 126-28
Foreman, Dave, 341
Forest City, Maine, 137-38
Forest for the Trees (Milliken), 82
Forest Legacy program, 312
Forest Resources of the United States (U.S.
  Forest Service), 129
Foss, Carol, 36-47
Fox, Stephen C., 121
French, Dale, 270-71, 291
French, Jeris, 270-71, 291
French-Indian War, 178
Future of the Adirondack Park, The
  (Temporary Study Commission on the
  Future of the Adirondacks), 277

Gadzik, Chuck, 73-92, 149-159
Georgia-Pacific, *xxiii*, 75, 133, 141-44,
  150-52, 157
Gerdts, Don, 288-91
Germany, logging and forestry techniques
  in, 91
Glens Falls, New York, 267
Goldsmith, Sir James, *xv*
Gorham, New Hampshire, 4
Governors Task Force on Northern Forest
  Lands, *xxii-xxiii*, 300, 305
Graham, Frank, Jr., *The Adirondack Park*,
  277
Grand Canyon, proposal to dam, 328
Grand Lake, 137, 141
Great Depression, 70, 177
Great Northern Paper Company, *xviii*,
  xxiii, 67, 75
Greene, Eila, *see* Moffat, Eila

Grenier, Paul, 308
Guarneri, Giuseppe, 247
Gulf and Western, 20

Hagenstein, Perry, *xx*
Haines, Sharon, 133-34
Halsey, Brent, 20-21
Halsey, Brent, Jr., 28
Hardwick, Vermont, 262
Harvard University, 82
Haynes, H.C. "Herb", 95, 145, 151-53,
  156
Hays, Samuel P., *Conservation and the
  Gospel of Efficiency*, 338
Hickey, Joe, 279-80
Holstein cow, 184-85
Home Depot, 215
Houston, Ned, 208-10
Howard, Bonnie, 271
Howard, David, 271
Hudson River, *xv*, 273, 274-75
Hudson River school (of landscape paint-
  ing), 273
Hunter, Malcolm, 41, 132
Hutchins, Curtis, 72
Hybsch, Mary Beth, 300

Inn on the Common (in Craftsbury), 203-10
Intercounty Legislative Committee (ILC)
  Planning Commission, 293-97
International Paper Corporation, *xviii*; advertisement in the *New Yorker*, 133;
  contracts with Treeline, Inc., 94-96, 99;
  employment of Brian Sowers, 93-94;
  forest management in Maine, 99-101,
  104-107, 115, 133-36; land ownership
  in Maine, 67, 100; manufacturing facilities in Maine, 100; participation in
  Maine "Forest Forum," 126; Bob
  Stegemann as official of, 268
Irish, Rick, 107-10
Irland, Lloyd, 128, 130-32

James River Corporation, contracts with
  Baskahegan Company, 75; demand for
  pulp, 324; land owned by, *xxi*, 30, 44,
  321; ownership of Berlin mill, 4, 20-26;
  relicensing of dams owned by, 308,
  334-35; sale of Groveton paper mill,
  24; sale of land for wildlife refuge, 11, 28

INDEX

Japan, as buyer of New England wood, 320

J.D. Irving, contracts with Baskahegan Company, 75

Jenkins, Warren, 35

Jersey cow, 184-85

John Hancock Financial Services, 321; *see also* John Hancock Timber Resource Group

John Hancock Timber Resource Group, 30, 44-45, 157, 321

Johnson, Charles, 302

Johnson, Vermont, 227

Johnston, Ted, 126-27, 304, 308, 314

Kelley, John "Dynamite," 68-71, 79, 80, 82, 152

Kennebec River, 64, 65

Kinney, Fred, 71-72

Kossuth, Maine, Baskahegan Company ownership in, 83

La Grasse, Carol, 267-71, 285

Lake Champlain, as shipping lane, 179

Lake George, New York, 276

Lake Ontario, *xiv*

Lake Placid Olympic Village, 289

Lake Umbagog, 3-14, 48-57

Lake Umbagog National Wildlife Refuge, 10-13, 26-30, 40, 340

Lake Umbagog Study Team, 28, 34

Lake Winnipesaukee, 27

Lamoille County Solid Waste District, 252

Land for Maine's Future Board, 28

Landvest, *xx-xxi*

Lansky, Mitch, 307; *Beyond the Beauty Strip*, 117

Last Stand (Manning), 123

League for Adirondack Citizens' Rights, 281-82

Leahy, U.S. Senator Patrick, *xxii,* 305

Leonard Pond, 11, 13-14, 28, 49, 51-54; bald eagle nest in, 27, 52-53

Levesque, Charles, 299-311, 317-18

L.L. Bean catalogs, paper for, 75

Local Government Review Board (LGRB), 278, 281-83, 289, 292-93

Loon Lake, New York, 280

Loon Preservation Committee, 31, 33

Lord (family): ownership in Maine, 65

Lyme Timber Company, 322

Lyndonville Teacher's College, 238

Machias River, as route for log drives, 68

MacNeal biomass power plant, 240-44

McPherson, Hank, 145

Mad River Valley, 322

Magalloway River, 3, 18, 40

Maine: early logging history, 61-67; early settlement of, 64-65, 68, 99-100

Maine Audubon Society, *see* Audubon Society of Maine

Maine Department of Environmental Protection, 110

Maine Environmental Priorities Steering Committee, 155

Maine Forest Practices Act, 127-28, 155

Maine Forest Products Council, 126-27, 155, 304, 314

Maine Forest Service, 119, 125, 127

Maine Department of Inland Fisheries and Wildlife, 110, 142, 144

Maine Land Use and Regulation Commission, 145, 155-56

Maine Tree Growth Law, 77

Maine Woods Reserve, proposed, 331

Maine, University of, at Orono, 36, 41, 78, 122, 130, 132, 193; Cooperative Forestry Research Unit at, 79

Manning, Bill, 193

Manning, Richard, *Last Stand*, 123

Maple Decline Project, *see* U.S.-Canadian International Maple Decline Project

Marble, Dana, 117, 149-51, 153-54, 156

Marvin, David, 169, 216, 225-30

Marvin, Lucy, 225-29

Maskell, Dave, 218-20, 223-24

Massachusetts Bay Colony, *xvi*

Massachusetts, Commonwealth of, early ownership of Maine, 64

Mattanawcook Pond, 93, 96, 110, 112

Mattawamkeag Stream, 62-65, 67

Mayflower, xvi

Merrimack River, *xv*

Merry Men (Chute), 250

Michener, James, *Texas*, 250

Milliken (family): 68-72, 76; *see also* Baskahegan Company

Milliken, Gerrish, 68, 70-72

Milliken, Roger, Jr., 77, 82, 126-27, 154-56, 314-15

Index

Milliken, Roger, Sr., 70-72, 80-81
Mitchell, U.S. Senator George, 304-305
Moffatt, Abby, 175
Moffatt, Andy, 181
Moffatt, Bob, 175-85, 199, 235, 254
Moffatt, Eila, 175, 177, 181
Moffatt, Jim, 163-263 passim
Moffatt, Larry, 169-70, 180, 181, 183, 212, 235-48
Moffatt, Mike, 235
Moffatt, Virginia, 181, 183
Montgomery School House Company, 227
Montpelier, Vermont, 230-32
Morgan, Ross, 211
Morse, Tom, 303-304
Mother Earth News, 192
Motyka, Conrad, 304
Muir, John, 121, 274
Murray, Rev. William H.H. "Adirondack," 273

Nash Stream, 37, 45
National Audubon Society, *xxii*, 286
National Environmental Policy Act, 338
National Forest system, 328, 338
National Park Service, 121, 340-41
National Park system, 328
National Wildlife Federation, 307
Natural Resource Council of Maine, 155, 306
Nature Conservancy, *xxi-xxii*
Neily, Sandy, 308-309
New Hampshire Department of Fish and Game, 12-13, 31, 33
New Hampshire Land Conservation Investment Program, 28
New Hampshire Timberland Owners Association, 300, 320, 339
New York City, as market for Christmas trees, 216-19, 233; business community's role in Adirondack Park, 274-76
New York State Department of Environmental Conservation, 268, 288
New York state legislature, 277, 284-86
New York Times, 123; editorial calling for creation of Adirondack park, 273-74; Sierra Club advertisement in, 328
New Yorker, International Paper advertisement in, 133
Newberry, Ruth, 281

Newcomb, New York, 270
Newport Furniture Company, 227
Niebling, Charles, 320-21, 324
North Country Council, 339
Northeast Logging, 151
Northern Forest Alliance, 306-308
Northern Forest Forum, 307
Northern Forest Lands Act, 305-306
Northern Forest Lands Council, *xxiii*, 156, 271, 299-319; Citizen Advisory Committees, 300, 306; "Finding Common Ground" (final recommendations), 310-18, 324; "Findings and Options," 302; listening sessions, 117, 267-71, 310-11; work affected by Adirondack dispute, 285
Northern Forest Lands Study, *xxiii*, 270, 305, 319; contained most steps recommended by Northern Forest Lands Council, 316; work tainted by Adirondack dispute, 285, 290
Noss, Reed, 131

Oak, Charles, 82
Oleson, Senator Otto, 28
OSHA (Occupational Safety and Health Administration), *see* U.S. Occupational Safety and Health Administration

Pacific Northwest "spotted owl" conflict, *xxii*, 320, 330-31
Parmachenee Lake, 3
Passamaquody Indians, 82
Patten Corporation Northeast, *xx*, 145
Penobscot River, *xv*, 63-65, 159
Philander (family): ownership in Maine, 65
Pinchot, Gifford, 121, 274
Pingree (family): ownership in Maine, 65, 133
Pingree, David, 65
Pingree, John, 133, 322
Prentiss (family): ownership in Maine, 65
Project SHARE (Salmon Habitat and River Enhancement), 157
Proulx, E. Annie, 259
Public Service of New Hampshire, relicensing of dams owned by, 334
Purdue, Richard "Dick," 281-82, 292-98
Putnam, Henry, 68

354

Rancourt Associates, *xxi*
Rangeley Lakes, 3, 27
Rapid River, 3, 18
Real Goods Toys, 227
Reidel, Carl, 305
Restore: The North Woods, 157, 307-308, 332-33, 339
Riendeau, Armand, 12, 33, 35, 49
Ripley, Arthur, 107-109
Roberge, Don, 4-57 passim
Roberge, Leo, 4-57 passim
Rockefeller Commission, 277-78, 285-86
Rockefeller, Governor Nelson, 277
Roosevelt, President Theodore, 121
Roucheford family, *see* Rushford family
Royal Navy, British, *xvii*
Rudman, U.S. Senator Warren, *xxii*, 305
Rushford family, 175

St. Croix River, 65, 68, 159
St. Regis Paper Company, *xviii*, 71
Samson and Adams (lumber company), 71
Sanderson, Byron, 88-89
Sayen, Jamie, 307, 308, 337
Schmitt, Michael, 202-10
Schmitt, Penny, 202-11
Seven Islands Land Company, 28, 87, 133, 322
Seymour, Robert, 130-32, 135
Shanck, John, 21-22
Sheehan, John, 292
Shorey, Tim, 97-109
Sierra Club, 328
Silvio Conte National Wildlife Refuge, 340
Slike, Dick, 95-99, 104, 111
Smith, Captain John, *xvi*
Smith, David, 72, 80-81
Society for the Protection of New Hampshire Forests, *xxi-xxii*, 306, 339
Society of American Foresters, 134, 155, 227
Soulé, Michael, 341-42
Souers, Brian, 93-116, 324
Spednik Lake, 137-39, 141-42, 146
Sports Illustrated, paper for, 75
Stafford, Senator Ronald, 282-83, 287, 290, 292-93, 295-96
Stange, Stephen, 304
Stegemann, Bob, 268-69, 308, 315
Sterling Academy, *see* Sterling College

Sterling College, 192-94, 199, 206-11, 260
Steward, John, 122-24
Stoddard, "Seneca" Ray, 273
Stowe Canoe and Showshoe Company, 227
Stradivarius, 247
Success Pond Road, 30-31
Sudharsan, Krishnan, 36-40
Swan, Henry, 321-22
Sweden, logging and forestry techniques in, 78-79, 91
Sweet, Barbara, 268-69
Syracuse University, 134

Taylor, Jeffrey, 21-22
Temporary Study Commission on the Future of the Adirondacks, *see* Rockefeller Commission
Texas (Michener), 251
Thibodeau, Paul, 32
Thirteen Mile Woods, 44-45
Time, paper for, 75
Topsfield, Maine, Baskahegan Company ownership in, 83
Tourin Musica, 227
Treeline, Inc., 93-116
Trust for New Hampshire Lands, 28, 300
Tug Hill Region, New York, *xiv*

Umbagog Waterfowlers Association, 12, 33, 49, 55
Umbagog, Lake, *see* Lake Umbagog,
Union Water Power Company, 26, 32-33
United Nations International Biosphere Reserve, *xv*
U.S. Army Corps of Engineers, 110
U.S. Bureau of Land Management, 327
U.S.–Canadian International Maple Decline Project, 168-69
U.S. Constitution, Fifth Amendment "takings" provision of, 282
U.S. Department of Agriculture, 121
U.S. Department of the Interior, 121
U.S. Farm Bill (of 1990), 300
U.S. Fish and Wildlife Service, 157, 333, 340; migratory bird censuses, 42-44; role in Umbagog National Wildlife Refuge, 11-12, 28-9, 34;
U.S. Forest Service, *xxii*, 121, 129-30, 300-301, 335

U.S. News and World Report, 204
U.S. Occupational Safety and Health
   Administration, 102-103, 115, 246
University of Maine at Orono, *see* Maine,
   University of
University of Vermont, *see* Vermont,
   University of,
Urie, Bruce, 196-98

Valmet Woodstar whole-tree harvester,
   85-88
Vermont: population, growth of, 189-90
Vermont Land Trust, 198
Vermont Natural Resources Council, 228,
   306, 339
Vermont Property Rights Center, 304
Vermont, University of, 164, 192, 305
Vietnam War, 188

Waddell, Michael, 334
Wagner Woodlands, 321-25, 328
Wal-Mart, 215
Ward, Andy, 111-116
Washington Post, Sierra Club advertise-
   ment in, 328

Waterhouse, Charlie, 218-20, 223-24
Wausau Paper Company, 24
Weeks Act, 338
Weyerhauser, Inc., 133
Wheaton, Dale, 137-48, 244
Wheaton, Jana, 140-41
Wheaton, Kim 141
Wheaton, Lance, 144
Wheaton, Ruth, 140, 142
Wheaton, Woodie, 140
Wheaton's Lodge, 137-148
White, Zebulon, 72
Whittaker, Brendan, 315
Whole Earth Catalog, 192
Wild and Scenic Rivers, 340-41
Wild Earth, 341-42
Wilderness Society, *xxii*, 286, 331-32
Wildlands Project, 341-42
Williams, Robert, 20-21
Woodworth, Neil, 268-69
Wyman, Brad, 28

Yale School of Forestry, 18, 72

DATE DUE